Unusual
Vegetables

Unusual Vegetables

Something New for This Year's Garden

by the Editors of Organic Gardening and Farming®

edited by **Anne Moyer Halpin**

 Rodale Press Emmaus, PA

Printed in the United States of America on recycled paper

Illustrations by Cynthia Hellyer
Book design by Ronald Dorfman

2 4 6 8 10 9 7 5 3 1

Library of Congress Cataloging in Publication Data

Main entry under title:

Unusual vegetables.

  Bibliography: p. 429
  Includes index.
  1. Vegetable gardening. 2. Organic gardening.
I. Halpin, Anne Moyer. II. Organic gardening and
farming.
SB324.3.U58   635   78-528
ISBN 0-87857-214-7

Contents

Acknowledgments

A sincere thank-you to all the organic gardeners who wrote in to share their experiences with growing unusual vegetables. Special thanks to John Meeker, James Jankowiak, Nancy Pierson Farris, Catharine O. Foster, Doc and Katy Abraham, Cathy Bauer, and Larry Korn for sharing the wisdom they've gained through years of gardening. Thanks also to Anna Carr for her editorial assistance.

—A.M.H.

Introduction

Try Something New in Your Garden This Year

BECAUSE gardening is a creative process, and gardeners are such tireless experimenters, the sense of excitement that comes from trying something new in the garden and the satisfaction gained from growing it successfully are constant incentives for gardeners to try their hand at new and unusual vegetables. Many organic gardeners have asked us where they can find the kind of information they need to expand their gardening experience and experiment with vegetables they've never grown before. They want to know about vegetables that will grow where more common types won't. They want to know how to grow vegetables that aren't sold in supermarkets. They ask questions like: What can I grow to extend my season? What can I grow in a hot climate, cool climate, dry climate?

Now that more and more people are discovering the rewards of gardening, there are lots of books available, some of which are very good, for the beginning gardener who wants to learn the basics of planning and planting a garden, making compost, using mulch, and controlling pests. And there are lots of cookbooks to tell you how to store and serve your harvest. But what about the gardener who's mastered the basics and is ready to try something new? Where can he or she go for information?

Some gardeners have simply begun to experiment on their own, but when they run into problems, they begin to ask questions, too. They want to know: Why did my Chinese cabbage fail? Why did my winter radishes turn out woody and spindly? Why didn't my celeriac germinate?

When we sought the answers to these questions, we found that information on growing and using out-of-the-ordinary vegetables was sparse and scattered. Botanical libraries had surprisingly little information about vegetables like martynia and tomatillo. Most of the books we did find were either aimed at the large-scale commercial producer, or were written by individual gardeners drawing primarily from their personal experiences in their particular parts of the country. Needless to say, an Iowa gardener's method of growing soybeans isn't going to be much help to someone in Vermont.

We decided the best way to answer all these questions would be to write our own book, to gather all the widespread information in one place. And we found that most of the information we needed did not come from books at all, but from gardeners who actually grow these vegetables.

There are people everywhere growing things their neighbors never heard of. On their own, through trial and error, they have found out how to successfully grow lots of vegetables you seldom find mentioned in books. It is to these veterans we turned for information and advice. Experienced gardeners can tell you things that seed catalogs don't; they can impart lessons that can only be learned by experimenting. We invited the gardeners who read and write for *Organic Gardening and Farming* to share with us their successes and failures. We asked several especially accomplished gardeners how they grow these unusual foods. As a result, you'll find throughout this book a collection of helpful tips, anecdotes, and personal experiences shared by gardeners who grow these vegetables. Their recommendations and explanations will give you a head start on growing these vegetables successfully.

The idea behind this book is not to provide you with eye-catchers or conversation pieces. There are some very practical, down-to-earth reasons for including some new and unfamiliar vegetables in your garden this year, and it's these you must consider when planning your plot. A gardener from Pasadena, Caroline Hoover, explains her family's philosophy this way:

We had great success in growing "far-out" vegetables over the last two or three years. The reason we grow them is because they add variety to our meals, are more nutritious than some of the store-bought vegetables, and also they are not available even in our local produce market. If they are available, the price is too high ($2.69 a pound for sugar peas). By growing the unusual, expensive ones ourselves, the only vegetables that we have to buy are the ordinary, less costly ones.

In the pages of this book you'll find 79 vegetables, arranged in 73 alphabetical entries. You'll probably come across some that you're already familiar with, and maybe even a few that you've grown at one time or another, but we're sure you'll discover some you never heard of or never thought to try. There are vegetables from other parts of the world—the Far East, Europe, and South America—that can be grown here in the United States; vegetables such as skirret and burnet, which gardeners of generations past grew and prized, but which have been largely forgotten today, displaced by more common but less interesting or less nutritious foods like parsnips or lettuce; ethnic and regional favorites, such as collards, which are popular in one part of the country, but which can be grown in other areas as well; plants like celtuce and nasturtium, which are actually two or more vegetables in one and have multiple uses; and vegetables which can be grown indoors on the windowsill or in the greenhouse to produce crops in winter, when fresh vegetables are few and far between. As you can see, these vegetables are not just novelties to grow for fun when you have extra space. They possess many valuable characteristics that make them worthy of a place in your garden.

Making vegetables interesting has traditionally been a problem for cooks, especially when children are involved. And even your favorites, however garden-fresh, begin to pale after many seasons of growing and eating. The vegetables in this book will add variety to your menus, without a doubt. But at the same time, they are not hard to use, and you don't have to learn a lot of new methods of cooking just to serve them. They can be substituted for many common vegetables in conventional recipes, to give your family's favorite dishes a new twist. These new vegetables will give you a chance to try your best bean soup with fava or horticultural beans; bring new life to your potato salad by adding some crunchy-sweet jicama; or use less expensive, better-tasting candied ground cherries instead of citron in your Christmas fruit cake this year. For a list of all the substitutions you can make, see the Quick Kitchen Guide in the Appendix section.

In addition to their exciting and unusual flavors, many of these vegetables possess extraordinary nutritional value. Did you know that dandelion leaves have six times the vitamin A and more than twice the calcium, iron, and phosphorus of lettuce? Or that Jerusalem artichokes and scorzonera store their sugar in the form of inulin rather than starch, and are thus a boon to diabetics as a source of carbohydrates?

Some of these vegetables are easier to store than conventional types. Coriander, for instance, can be frozen without blanching, and for many dishes is superior in this form to the dried herb. Winter radishes will keep throughout the winter in a root cellar or storage pit or barrel, but spring radishes stored this way soon develop hollow centers. Some vegetables, like celeriac, leeks, salsify, scorzonera, and Jerusalem artichokes, don't have to be harvested or stored at all! Even in New England and other areas where the winters are cold, you can dig these vegetables all winter long whenever you need them just by putting a thick layer of mulch over them to keep the ground from freezing solid.

A bit of careful planning can extend your growing season to let you make optimum use of your garden space. Vegetables like broccoli raab, garden cress, and daylily will give you fresh vegetables in early spring—way ahead of your radish or lettuce crop. On the other hand, collards, kale, Hamburg parsley, and corn salad withstand frost to give you fresh vegetables well into autumn. If you live where climatic conditions are too extreme for your favorite vegetables, you may find that there is a tasty substitute especially suited to your kind of weather. Basella and New Zealand spinach both thrive in hot weather that causes regular spinach to bolt. Celeriac loves a cool climate where it's difficult if not impossible to grow celery (and many folks find they like its mellow, celery-like taste even better than celery).

If you are plagued with poor soil that's always made gardening a problem, or if you just don't have the time to do a lot of fertilizing, you should find success with crops like borage, purslane, nasturtium, and Jerusalem artichokes. Some of these foods, such as burdock, purslane, and dandelion, may sound like common weeds to you. But these are more than just weeds; they are improved varieties which have been domesticated for garden growing. Domestic dandelion and purslane have bigger, better-tasting leaves than their wild cousins. Edible burdock develops huge, fleshy roots, and is harvested before it even gets the chance to form the pesky burrs that are so annoying in the wild plant. There are two vegetables in this book, nettle and Good King Henry, that are still considered wild plants because they have not yet been domesticated for gardens, but are nonetheless valuable enough to merit a spot in the garden. For the others, domestication has capitalized on the assets of these food plants, and they can now be considered full-fledged garden crops.

In our coverage of these extraordinary vegetables, we've tried to be as complete as possible in order to make this book a useful reference. There's a bit of the history and folklore surrounding the vegetables, and a description of each plant's growing habits and climatic preferences. Instructions for growing the vegetable follow, including when and where to plant, what type of soil is best and how to prepare it, moisture needs, and extra care that will help to insure a superior crop. Pests and diseases to which the plant is subject are discussed, and remedies are recommended. You'll find directions for how and when to harvest your crop, along with suggestions for storage. Finally, the vegetable's value as a food and its place in the diet are assessed, and ideas for preparation and serving are presented. There are even some recipes.

Sources of seed and, in some cases, plants, are listed for every vegetable. More information on the companies listed in each section may be found in an appendix which provides addresses and brief descriptions of each firm's specialties and orientation.

Whether you're a veteran gardener or a novice, you will, we hope, find all the information you need to grow all these special vegetables. On, then, to see what can be new in your garden this year.

Unusual Vegetables

Amaranth

Amaranthus, spp.

ONCE the sacred food of the Aztecs, the majestic amaranth seems destined for a future as noble as its past, for this statuesque plant possesses staggering potential as a first-rate source of both grain and leaf protein. Found on five of the seven continents, the many black- or tan-seeded species of amaranth take their name from the Greek word that means "unfading," and through history these magenta-tinged plants have been a symbol of unending life. Ancient artists used their image to decorate tombs and likenesses of the gods, and poets employed it to signify an imaginary flower that would never fade.

Originating over 6,000 years ago, the first amaranths were drought-resistant plants that took readily to full sun and disturbed soil. Spreading rapidly in naturally open sites, the rugged plant suffered little competition from other plant species. Thus the amaranths had established themselves throughout the temperate and tropical world before man played any role in their development. Later, they apparently followed human communities, taking root easily in the soil broken by primitive peoples. First gathered wild, little by little these omnipresent plants were deliberately grown, then bred for better leaves or larger yields of lighter-seeded grain.

A host of black-seeded vegetable amaranths have been grown throughout east Asia since ancient times for their

1

flavorful leaves. Today this hot-weather version of spinach is one of the most popular potherbs in tropical and subtropical parts of the world, while the redroot pigweed native to the United States remains one of the most foraged wild plants in this country. The so-called "grain amaranths" with larger, ivory-colored seeds supposedly were first domesticated in North America over 4,000 years ago, evolving from a primitive "pioneer species" growing in deserts and canyons. There is evidence that grain amaranth persisted as a crop among the Paiute Indians of our Southwest as late as 1870. But the most dramatic and economically important use of this grain in the Americas so far seems to have taken place in Mexico, where the colorful plant was domesticated by the Aztecs long before the Spanish Conquest. When Cortés arrived in 1519, amaranth was almost as important as corn. Montezuma received 200,000 bushels a year as tribute, and the small ivory-hued seeds played a major role in the Indians' religious festivals. Mixed with honey—and sometimes with human blood—the seeds were kneaded into a paste called *zoale.* This was shaped into idols that were paraded through the streets, then broken and eaten by the celebrants. Because amaranth was so closely linked with such pagan rituals, the Christian Spaniards suppressed it as a crop, and today the nearly extinct grain is grown in only three areas of Mexico. Ironically enough, it is used mostly to make a popular candy reminiscent of *zoale*!

For the last century or so, amaranth has been a significant grain crop only among Asian hill tribes, though it is now spreading into the plains of India, where in some places it outyields adjacent stands of corn. During the last few years, this forgotten food has also been grown in the United States by over 13,000 *Organic Gardening and Farming* reader/researchers. Their respectable yields in diverse places from seed bred in Mexico confirms that amaranth is a remarkably adaptive plant that utilizes water, light, and atmospheric carbon with extraordinary efficiency. This ability helps amaranth to adjust beautifully to both the warmth of Mexico and to the snow-swept heights of the Himalayas, where it is grown in places too high for permanent human habitation.

• Not only are we enjoying the greens (at all stages—even now with the plants 50 to 60 inches tall) but so has everyone with whom we have shared them. My husband especially likes the stems which, so far, have cooked up nice and tender. His mother, a greens eater in her eighties, says that they are the best greens that she has ever eaten and also appreciates the ease with which they can be prepared.

—*Mrs. Ray R. Fox*
Cheektowaga, New York

Taking about 120 days to mature seed in southeastern Pennsylvania, amaranth is an annual herb with relatively small leaves that vary from linear to nearly wedge-shaped and range in color from dark green to magenta. The plant is upright and sparsely branched, and has the curious ability to flower at almost any size and age if day-length conditions are favorable. Moreover, it flowers abundantly, sending forth small blooms in reddish terminal or axillary spikes. Though Chinese spinach types grow only 2 or 3 feet tall, the grain amaranths can be spectacularly showy, reaching well above human height and displaying richly extravagant flower heads up to a foot long and 5 inches wide.

The various amaranths under cultivation include ornamentals that have been part of the European gardening scene from at least the early eighteenth century. Invariably dark-seeded and with reddish leaves, these species are offered in gardening catalogs as love-lies-bleeding and Prince's-feather. One of the tastiest amaranths grown for greens is *A. gangeticus viridis.* Known as *tampala* in India, this variety was introduced to Americans in the mid-1940s by the Burpee seed company, from a strain a retired missionary brought back from China. Of the three major species of grain amaranths, the most widely grown is *A. hypochondriacus,* which is currently found in Mexico, Guatemala, Iran, India, Afghanistan, the Himalayas, parts of China—and in the gardens of Rodale Press reader/researchers. The flower heads of this amaranth are long, thick, and often erect, and in growth trials in Pennsylvania this species significantly outyielded *A. cruentus,* producing a pound of nutritious seeds per square yard of planting, plus a lot of small lower leaves with a beety taste that were harvested through the season.

If you're in a temperate clime and want amaranth to go a full season so you can harvest seed, you can direct-sow quite early in the spring (or, in the subtropics, in September). Amaranth germinates well even under cold and somewhat dry conditions, though the plant grows much faster when the weather is warm, flourishing best when the days are hot and the nights brisk. If your growing season is quite short, you might want to start grain plants early indoors or in a greenhouse. But you might have to transplant them several times before planting out. Performing decently even in drought-stricken areas, amaranth will make the best of poor and dry soil. But for optimum results, try well-tilled soil with plenty of manure or compost worked in, for amaranth has a big appetite for phosphorus and for nitrogen, which is known to increase the succulence and protein content of the leaves. If you like, broadcast the seed and use the thinnings for potherbs, leaving about 10 inches on all sides of grain plants. (Some researchers say that closer spacing of 8 inches by 8 inches stimulates the height of plants and increases yield per unit area.) Or you can sow thickly in rows 6 inches apart and ¼ inch deep, thinning to 2 inches apart when 2 inches tall and—if a grain type—to 8 or 10 inches apart when 8 inches tall.

If you have vegetable amaranth and plan to harvest it in one fell swoop when it's from 6 to 10 inches tall, you might want to do some succession plantings every 10 days to two weeks to assure yourself of a supply well into fall. On the other hand, your first planting will yield from April to October if you prevent flowering and prolong the life of the plants by harvesting 4- or 5-inch tips at two- to three-week intervals. (Interestingly, the iron content of the greens nearly doubles from the seventh to the ninth weeks, so you might want to delay harvesting the leaves of at least some of your plants.) You can store, can, or freeze extra leaves as you would spinach—or use them to prepare a leaf protein concentrate that will keep for two weeks in the refrigerator (see box).

Amaranth makes a good companion plant for potatoes, onions, and corn because it pumps nutrients from the subsoil. As your crop grows, it's probably best to provide moderate amounts of water (if needed), to mulch thickly, and perhaps to add fertilizer a couple of times, though some folks growing grain amaranth claim that plants they leave unwatered, unmulched, and unfertilized outperform those that are pampered. Since chewing insects like amaranth leaves, you might want to protect the amaranth bed with fine screening. If you're growing grain amaranth, you should also keep your garden well weeded, for the plants have a tendency to hybridize with pigweed. The root system of amaranth tends to be weak, and as the plants grow heavy, you might give them support to keep them from being beaten down by strong winds and rainstorms.

Harvest the huge seedheads when the seed begins to ripen and fall, and be prepared to lose some, for the flower heads mature progressively, and you will probably notice some shattering of the early ripening portions—a problem now being tackled by those doing selective breeding. Thresh and winnow by hand, which is easier than you might think because amaranth has one-seeded capsules that split along a natural line.

Nutritionally, both the leaves and grain of amaranth are of unusual value. Tasting like spinach with a touch of horseradish, the raw greens have substantially more calcium than beet greens, kale, chard, and spinach, and more iron than all these leaf vegetables and collards as well. Because the gigantic amaranth yields four times more green matter than comparably light-and-carbon-dioxide-efficient plants, researchers have declared it an outstanding source of leaf protein concentrates that can be used as fodder or as human food. Amaranth grain is also a nutritional bonanza, for it contains high-quality starch and more protein than any other grain: over 15 percent. What's more, that protein comes closer to meeting the ideal requirements of the human body than cow's milk (measured against the ideal protein rated 100, amaranth scores 75, versus 56.9 for whole wheat, 68 for soybeans, and 72.2 for cow's milk). Amaranth seed is also most unusual and valuable because it is high in lysine, the amino acid most lacking in cereals such as corn, wheat, sorghum, and barley. This means that relatively small amounts of nutty-tasting amaranth mixed with these grains in muffins, breads, pancakes, and other baked goods can provide complete protein. Thanks to the unusual amino acid profile of amaranth, it can also complete the protein present in beans.

You can eat the stems and leaves of young stem tips together, but it's best to cook the more mature stems alone for 8 to 10 minutes (they taste a little like artichokes). To retain the

• One day our daughter and her husband ate with us when we had amaranth cooked and served with hot cooked bacon dressing. We all liked it very much. A few weeks later they dropped by, and our son-in-law was wondering if we were having any more of that "good stuff" for supper. We've also tried cooking it with homemade spaghetti sauce, rice, and a few slices of left-over meatloaf chopped in it. That is good too.

—*Mrs. Clark R. Minnich*
Boiling Springs, Pennsylvania

most iron and vitamin C in the leaves, steam them for 10 minutes. You can then serve them with butter or mixed with peanut butter you've blended with water. If you like, add the raw leaves to soup broth or stir-fry them in heated oil in which you've browned a garlic clove, then stir-fry in some ground pork, add boiling water, and simmer a while. You can also chop and stir-fry the larger shoots with bean sprouts and other vegetables, adding a little soy sauce and water during the last 3

• We found the growth rate of the amaranth to be amazing. Our neighbor swore he could see it growing from his window. Found the greens to be so tasty. Had a heavy windstorm which blew over and broke two plants, which we fed to our quarterhorses. They ate it all down to about 1 inch from the roots, and loved it. By the way, we laid a corn stalk next to the amaranth stalk in front of them, and they picked amaranth first. The corn died from not enough rain, but amaranth grew on.

—*Will and Nancy Ackerman*
Fayetteville, Tennessee

AMARANTH: TAKE IT OR LEAF IT

Though green leaves are the earth's most plentiful source of protein, they also contain lots of water and fiber, and have a protein-binding structure. All of this means that it's impossible for humans to get a substantial amount of protein from leaves eaten straight off the plant. Researchers have found, though, that when leaves have at least 76 percent water and at least 18 percent protein on a dry weight basis, as the amaranths do, it is possible and worthwhile to make them into a concentrated food tremendously rich in protein, vitamin A, calcium, and iron. This is done by breaking down the structure and fiber of the leaves and then pressing out the water. The pleasantly aromatic, mild-tasting leaf nutrient concentrate (LNC) gotten this way has a crumbly texture much like dry cottage cheese. Although it takes rather a lot of effort to produce LNC in sizeable amounts, the product is very versatile, and may prove especially useful to vegetarians who eat no animal products whatever. Containing up to 65 percent protein, LNC may be blended into stews, sauces, mashed potatoes, and many other dishes.

Amaranth yields a more palatable leaf concentrate than alfalfa or soybeans, so if you are a culinary adventurer, you might want to try converting your mature amaranth leaves into LNC. For best results, harvest the leaves before the plants flower, for those gathered later when you harvest the grain are to 5 minutes of cooking. Or try incorporating the leaves in vegetable curries as people do in India and Ceylon. For more good eating, combine a pound of cooked, drained fresh amaranth with 1 pound of ricotta cheese, 1 beaten egg, and ¼ cup grated Parmesan, and bake at 350°F. for 30 minutes. Or try raw amaranth chopped and mixed with chopped onion, slightly beaten eggs, and a little salt, then fried as small pancakes in safflower oil. This green also tastes great when it's cooked and

likely to be bitter.

The yield of this recipe is low, and for regular use you will need to multiply it. But for the purpose of experiment, you'll want to start small. To make 3 grams of this new food, gradually add ¼ pound of washed amaranth leaves to 1 pint of water in a blender, liquefying as you go. (If your water tends to be alkaline, add a little vinegar.) When all the leaves have been added, continue to blend at high speed for another 2 minutes. Then pour the mixture through a fine cloth spread over a colander or sieve, catching the liquid and squeezing the cloth to get as much of it as possible. Next, heat this liquid quickly to 176°F. (80°C.), stirring all the while. When the mixture reaches the right temperature, allow it to stand, off the heat, until the solid protein starts to separate from the liquid. Then pour the mixture through a fine cloth, allowing the liquid to drain off, and squeeze the solid protein lightly. While the curd is still in the cloth (in the colander), rinse it with water, stirring lightly. Allow it to drain again, then squeeze it with your hands or place the bag in a potato ricer. When you remove the pressed curd from the cloth, you can use it right away, freeze it, or store it for up to two weeks in the refrigerator. You can sneak this green protein into Greek spinach pie or a vegetable casserole. Or add it to peanut or sesame butter and roll into small balls to eat as a snack.

then added to seasoned tomato sauce. If you like, you can mix the cooked greens in a blender with minced garlic, parsley, basil, oregano, tomato sauce, and some tomato paste, then use this mixture as one of the layers in a lasagna that also features broad whole wheat noodles and a blend of ricotta cheese, salt, pepper, and parsley.

You can prepare your amaranth grain most tastily by using it in your favorite whole wheat bread or muffin recipe in the ratio of one part amaranth flour to four parts whole wheat. Amaranth milled in a blender or burr mill also can be made into tortillas. And toasted, milled amaranth can be added to cooked cereal. If you pop the seeds in a dry skillet or wok, you can mix them with honey to make a tasty candy like those enjoyed in Mexico and India. Or you can powder the toasted seeds and mix them with water to make the Aztecs' favorite drink. To make hearty mixed grain pancakes, mix together thoroughly ½ cup whole wheat flour, ¼ cup cornmeal, ¼ cup amaranth flour, 1 egg, 1 cup buttermilk, 2 tablespoons oil, 1 tablespoon honey, 1 teaspoon baking powder, ½ teaspoon baking soda, and ½ teaspoon salt. Pour the batter by small portions onto a hot oiled pan or griddle, and cook until brown on both sides.

You can order various amaranths yielding especially tasty leaves or large-seeded grain from Burgess Seed and Plant Co. (Tampala); W. Atlee Burpee Co. (Fordhook Tampala); Grace's Gardens (Chinese spinach, Hinn Choy); Gurney Seed and Nursery Co. (grain amaranth); Le Jardin du Gourmet; Tsang and Ma (Chinese spinach, Hinn Choy); and Thompson and Morgan, Inc. (Hinn Choy).

Bamboo

Phyllostachys, spp.

MOST Americans no doubt think of bamboo as an Asiatic plant, and a tropical Asiatic plant at that. But bamboo is able to withstand cold weather with little problem, and it can be grown in many parts of the United States. Bamboo has been known to grow wild in parts of the Western hemisphere, although the clearing of lands for agriculture has obliterated much of the wild bamboo. While not truly a garden plant, its value as a food, along with its host of other uses and the graceful, stately appearance of the full-grown "reeds" make bamboo well deserving of a place in your yard or garden.

Bamboo is one of those plants that has a thousand-and-one uses that make it invaluable to the people in regions where it grows abundantly. The Japanese, for example, depend on it heavily. Although the people do not cultivate it specifically for its shoots, most farm households have a bamboo grove very near the house. If not, there is almost certain to be a grove used and managed cooperatively by the village. In addition to producing shoots for food, bamboo is also used in the Orient in constructing the walls and roofs of houses, for room dividers, baskets and furniture. Over the centuries, bamboo has come to symbolize long life, straightforwardness, and the attributes of tranquility, peace, and modesty which we in the West have learned to associate with the Eastern way of life.

11

You might expect that a tree-like plant associated with an attitude of tranquility would grow like the redwoods in California at a steady but leisurely pace, taking many years to mature. Not so the bamboo—its rate of growth is astounding. Depending upon the species, bamboo can attain a height of 120 feet, and can grow 30 feet or more in a single season! A California gardener who lived in Japan for several years once measured a MIMOSA bamboo to grow 13 inches in a 24-hour period during the monsoon rains there. Bamboo may look like a tree, but it is actually a primitive kind of grass. The part of the plant which is eaten, the new shoots, develop from the underground root system, and are protected by tough, tightly fitting, overlapping husks. There are hundreds of varieties of bamboo, but the slender ornamental type seems to be most often cultivated in this country. However, there are several edible varieties that can be grown in the United States.

The type most often grown for its shoots, MOSO (*Phyllostachys moso*), is tall and most impressive to see in a backyard. It will often attain a height of 80 feet, with rather wide-spreading branches. The shoots are quite delicately flavored. It's hard to believe, but Moso shoots can grow at a speed of 2 inches an hour. Moso is widely grown in Japan, sometimes on huge plantations. Commercial growers often cut off the tops of the plants when they get 12 feet high, so that the branches form a leafy canopy. The result is a larger crop of small shoots—just what the grower wants.

Another excellent edible variety is called SWEET-SHOOT (*Phyllostachys dulcis*). It is medium-large in size, and is an early-sprouting type. Sweet-Shoot is said to be the type of edible bamboo commonly grown in central China. The shoots are tender and, as their name implies, very sweet. GREEN SULPHUR (*Phyllostachys sulphurea viridis*) is yet another bamboo of good edible quality, and is noted for its delicate fragrance.

In order to grow well, bamboo has just two basic requirements—good drainage and an acid soil. Plant bamboo in holes about 2 feet wide, and cover the rhizomes with about an inch of soil and work in some organic fertilizer. In the year after planting, an application of a high-nitrogen fertilizer, such as blood meal, will be helpful. Cultivate lightly during the first year; after that all you need to do is keep the taller weeds trimmed.

The rhizomes should be planted at least 6 feet apart. If you're growing the larger Moso variety, space the rhizomes 10 to 15 feet apart.

Kathleen and Al Kaule, two Arkansas gardeners who have been growing bamboo for the past several years, report that from the start they have found it always easy to grow and very hardy—their bamboo has survived even ice storms. Although the Kaules have seen bamboo thriving in full sun, they've found that theirs does best in partial shade. Bamboo is generally pest-free.

The shoots are most tender if they are cut as soon as they break through the ground. In any event, they must be gathered before they grow too large. Once the sprout has grown more than about 2 feet tall, it becomes woody and inedible. Sometimes the shoots are blanched by mounding up dirt around them, in the same way asparagus is blanched, to get longer shoots. An established stand of bamboo will provide plenty of shoots for eating and still leave enough "trees" to lend a gracefully exotic appearance to your yard or garden. Indeed, it's very sound practice to harvest some of the new shoots as they come up in spring, even when bamboo is grown primarily for ornamental purposes. If all the new shoots are allowed to grow to maturity, the grove will soon become too crowded. Disease and insects then become a problem, and the quality of the bamboo "wood" deteriorates.

To harvest bamboo shoots, dig a hole and cut the shoots several inches below the ground surface. The shoots spoil quickly once they are cut, and should be boiled and eaten the day they are gathered, if possible. If you must store them, you might try drying the shoots the way the Japanese do. Boil the shoots for at least 15 minutes after removing the husks, then salt the shoots and set them out to dry.

When preparing bamboo shoots for cooking, cut off and discard a small section of the base. Then peel off the other husks, and cut the sprouts into thin slices. It's not widely recommended to use bamboo shoots raw in salads, because some varieties have an unpleasant bitter taste in the raw state. Boiling the shoots for about 20 minutes, with a change of water after the first 10 minutes, will eliminate any bitterness. The Sweet-Shoot and Green Sulphur varieties are reputedly free of any bitterness, but to be on the safe side, you can cook the shoots and chill them for use in salads.

Bamboo's crunchy texture and mild flavor (somewhat reminiscent of young field corn) make it amazingly versatile in the kitchen. In addition to its standard use as an ingredient in oriental stir-fry dishes, bamboo shoots can be included in soups, pan-fried in *miso* (a salty paste made from a mixture of fermented rice, soybeans, and salt, that is used extensively to flavor Japanese dishes), incorporated into "American-style" dishes such as potato-bamboo salad or bamboo creole, and pickled or candied. Outside the kitchen, bamboo has lots of other uses. The leaves make good supplemental winter forage for livestock. The poles make excellent garden stakes, chicken fences, and tree props. The extra long poles can be made into handles for pickers to harvest fruits and nuts from trees (metal baskets for these pickers are available from seed companies, but the poles are not included).

Edible bamboo is difficult to find. One source is the nursery department at Bill Boatman and Co., South Maple Street, Bainbridge, Ohio 45612, which stocks it in spring. Another supplier is Lakeland Nurseries. You might also contact the nurseries in your area to see if they stock it.

In Japan, bamboo shoots and other wild mountain vegetables such as colt's foot, watercress, bracken fern and sword fern
shoots, wild onions, and mugwort are the first vegetables which
can be eaten in spring. During the months of March and April,
before the garden vegetables are ready to be harvested, these
wild plants traditionally make up the main vegetable diet of the
Japanese people.

Basella

Basella alba; B. rubra

BASELLA is known variously across the world as Ceylon Spinach, Country Spinach, Libato, Malabar Nightshade, Malabar Spinach, Vine Spinach, *Pasali*, and *Pu-tin-choi*. As the names suggest, it is primarily an Eastern plant, native to India and the tropical regions of the Far East. Gardeners and farmers in these countries cultivate it as both an ornamental and a vegetable perennial. In temperate regions it can be grown as an annual, warm-weather substitute for spinach. The thick, succulent leaves have a flavor much milder than that of chard or turnip tops. Being smooth, they are easy to clean and prepare for table. And because the plant is a climbing vine, the leaves are never gritty like spinach.

The plant belongs to its own family, *Basellaceae*, a little-known group which shares many of the characteristics of the beet and spinach family. Although there is only one species of basella, two names are often applied. The common *B. alba* has dark green, shiny leaves on vines that grow up to 4 feet in length. Leaves are round or oval-shaped and very thick. Small white flowers borne on spikes give way to round, purplish fruits. *B. rubra* differs from *B. alba* only in that its leaves, stems, and flowers are slightly tinged with red or purple. The color disappears in cooking.

As a tropical plant, basella does not withstand frosts and

grows poorly, if at all, in cold weather. Night temperatures of above 58°F. are required for seed to germinate, and the best results are obtained when they are well above that. But, since the plants do not require a long season, the crop can be easily raised in most temperate regions of the United States and southern Canada. Direct-seed in the garden well after danger of frost has passed. Sow seeds 1 inch deep and ¾ inch apart, in rows 2 to 3 feet apart. Later, after the plants have become established, thin them to stand at 1-foot intervals.

To obtain a lush crop where the season is shorter, start seeds indoors very early in the spring. Sow them in individual pots, eight to ten weeks before the last frost is expected; when the soil has become very warm and the weather mild enough for eggplant and melon, transplant the young basella to the garden bed.

Plants can also be propagated from slips or stem cuttings. If care is taken to shade the vines and provide plenty of moisture, they will root quickly.

Basella tolerates many soils, but prefers a sandy loam with a pH of 6.0 to 6.7. In tropical regions where the soil is moist and rich, seeds germinate within a few days; elsewhere two or three weeks may be required. Plants respond very favorably to applications of organic nitrogen fertilizers and frequent waterings.

If given supports on which to climb, basella plants will produce clean, grit-free leaves all summer until autumn frosts strike. In very warm regions or in the greenhouse, basella can be grown as a perennial, and will maintain itself for several years and produce continuously.

The first harvest can be taken within a month after planting, but this often results in stunted plants. It is best to wait until each plant has many leaves and, until the plants have branched, to pick only one leaf per vine. Later, large harvests can be taken. After about three months, the vines should be pruned every week or so to encourage the continued growth of young, succulent leaves.

To harvest, simply cut the tips of the vines, about 3 to 5 inches long. Basella branches readily, and frequent cuttings

• Some gardeners have found basella so popular with their families that they grow it as their primary greens crop. One gardener reported that: "Summer is no longer spinach-doldrum time in our house! Last summer we grew Malabar Spinach, and we intend to plant three times as much this year, in order to have frozen greens all next winter. If your family dislikes the strong taste of cooked chard, turnip tops, and other summer greens, you too will welcome this delicately flavored oriental vegetable."

help to keep the plant under control. Perennial plantings should be extensively pruned during flowering season to encourage and maintain the production of new, green shoots.

Basella has been judged relatively free of insect or disease problems, but there is one pest which sometimes attacks the plant—a fungus, *Cercospora beticola,* causes spotting of the leaves and eventually makes holes in them. The best defense is to pick off affected leaves as they appear and destroy them. Ordinary garden sanitation and crop rotation also help to discourage the spread of this fungus. Beets are also susceptible to this disease, so avoid growing Malabar Spinach near beets.

A planting of healthy, lush basella vines provides delicious greens for summer use and for freezing. Basella leaves are rich in vitamins A and C, and are a good source of calcium and iron as well. Because of their heavy substance, fewer leaves are needed to make a meal than with most greens. Oriental cooks prepare both leaves and stems as a potherb, steaming or stir-frying them. The vegetable can also be added to fish or shrimp for a light soup, or mixed with stew or other vegetables. Try it in your favorite spinach recipes—creamed, or au gratin, or with mushrooms, for example. Flowers, unless very young, are tough and undesirable fare and very old leaves are also to be avoided. When cooked, the green variety of basella retains its deep green color. *B. rubra* loses much of its pigment when cooked, and is generally considered less attractive than the green form. Although the odor of the cooking leaves is rather strong, the leaves themselves have a mild, delicate flavor. Take care not to overcook them, or the stems will become somewhat gelatin-like.

Seeds of basella can be purchased from W. Atlee Burpee Co. (under the name of Malabar Spinach), Gurney Seed and Nursery Co. (who call it Climbing Spinach), Nichols Garden Nursery (Malabar Spinach) and Glecklers Seedmen, who carry both the green and red varieties.

Asparagus Bean

Vigna sesquipedalis

MOST widely known as the yard-long bean, this long, long green bean is not really a bean at all—it belongs to the same family as the cowpea. The asparagus bean is widely grown in Asia, and has long been raised by European gardeners as well. Although seldom seen in grocery stores, asparagus beans are as good to eat as they are strange to behold, and make an interesting and welcome addition to any vegetable garden. They are easy to grow, produce abundantly, and have a pleasing taste all their own. Some people find that the flavor, true to its name, reminds them of asparagus. Others do not, but like the taste anyway.

The first thing to consider when planting asparagus beans is their love of climbing. Yard-long beans grow up, twine and twist, holding onto any vertical means of support they can wrap themselves around. They will do well staked, and even better when they can climb up strings or wires. Heavy twine laced between wires provides an excellent trellis. The plants will also be happy with 7- to 10-foot tall tripods to climb. A tripod is simply three long poles bound together at the top with wire, and arranged with their bases about 2 feet apart.

The plants themselves thrive in hot weather. As the season progresses, they turn into bushy vines that have been known to overgrow everything in sight during a summer hot spell,

• I planted a few of the small bean seeds in some starting mix. They germinated quite fast and put out a second set of leaves in no time at all. By April the rest of my seedlings were ready to transplant to the garden.

They looked awful silly, 3-inch plants growing around the base of a 7-foot bamboo tripod. By June the rest of my garden was going full steam ahead. The peppers were blooming; the tomatoes were beginning to set, but my tropical beans had hardly grown more than a few inches. I consoled myself by reasoning that tropical beans ought to grow in the tropics, not the hot dry heat of the central valley of Northern California. About the time I was ready to give up all hope of those delicious beans ever amounting to anything, our summer heat wave hit. With the temperatures hitting 95 to 100 degrees, those beans really took off. They began to climb all over everything, including a nearby apricot tree; not at all particular about climbing the stakes I had put for them, and blossoming as they went.

—*David P. Sauder*
Paradise, California

21

• We have been growing asparagus beans for three years. The beans we originally planted came in a package with instructions in both English and Japanese; the label stated they were Japanese asparagus beans. We bought them in desperation because all the local stores and nurseries were out of ordinary green beans by late June.

Desperation turned to delight when our seeds turned into tall bean stalks. We went on vacation for one week when the beans were not quite large enough to pick. On our return, we found the individual beans had grown to over a foot long! Without giving it a second thought, I had the water simmering, the beans cut up and in the pot. I cooked them until just tender and served them with salt and butter. They tasted almost identical to pole beans—my husband swears he can't tell any difference. To me, they have a zippier taste. Since then, I have used them in all my recipes that call for green beans.

—*Marcia Ambler*
Los Alamitos, California

including nearby trees. Although the vines can seem to be endless, their foliage, while abundant, isn't quite as thick and leafy as that of other beans. The plants develop very tenacious root systems. One New Jersey gardener remembers well the time that, "along came Hurricane Belle one day, and blew down all my pole bean teepees, except for the teepees anchored steel tight by the asparagus beans' three-quarter-inch rootstock."

Asparagus beans should be planted in spring, after the ground has warmed, the same time you probably plant the rest of your pole beans. If you've chosen to construct tripods for the plants to climb, plant the seed 6 to a pole at the base of each pole, then thin to 3 plants per pole when the plants are a few inches high. If you are planting in a row, along strings or a trellis, sow the seed 8 inches apart and 2 inches deep (less if your soil is heavy).* In order to have a good supply of asparagus beans for a family of four, you'll need a row that's at least 15 feet long. Some gardeners like to make two or three successive plantings in a long season, to insure a continuous crop.

Generally speaking, yard-long beans respond to the same kind of care that is given to regular pole snap beans. A rich, friable garden soil not too high in nitrogen is right up their alley. They are heat resistant and will tolerate some dryness, but they give better results with normal watering. A good rule of thumb is to see that the plants get water once a week. If rainfall isn't sufficient, they will benefit from some added water.

The vines form pretty white or purple flowers, each of which will produce a pair of pods (as do peas). As the beans are picked, new ones will grow from the same nodes, and the plants will keep on producing all summer long.

If left to their own devices, asparagus beans will grow to be as much as 24 inches long, and will get about as thick as a regular snap bean. Mark Demitroff of New Jersey reports that when the beans reach this stage, "it is always fun to tell the neighbors about how I sold the cow for some organic bean

---

* The planting depths listed throughout this book are those generally recommended. Gardeners with heavy or clay soils, however, may have problems with germination, and generally have better luck by just barely covering their seeds.

seed." Unfortunately, that's about all your beans will be good for if you allow them to mature. If not picked when tender, those long pods will become just as tough as any other bean or pea past its prime. Pick your beans before the seeds fill out the pods, while they're still crisp and tender. The pods will be about a foot and a half long, and roughly half the thickness of ordinary snap beans. Length varies, of course, so always use tenderness as your guide to harvest time.

Most people report few if any pest problems with asparagus beans. In fact, some say they are downright indestructible. One gardener reports that when his other beans are covered with Japanese beetles, the asparagus beans are spotless. Neither foliage nor beans show any sign of attack.

After you've successfully raised these strange-looking beans, how do you eat them? First of all, don't be surprised to find them on the menu at your favorite Chinese restaurant. They go well in stir-fried dishes mixed with meat, sprouts, soy sauce and seasonings. But they make an equally tasty side dish, and you can prepare them in the same ways you'd cook ordinary snap beans. Try them in a snap bean salad. Chop them into bite-sized pieces, and stir-fry or steam briefly. Don't expect your asparagus beans to look, taste, or smell like regular snap beans, though. They are more slender, denser, and less juicy than green beans. Their unique pea/bean flavor is difficult to describe, but words like subtle, zippy, or nutty have been used by those who find that the beans don't taste like asparagus. The leaves and stems, when young, can also be steamed for a good-tasting green vegetable.

Sources of seed for asparagus beans include: Burgess Seed and Plant Co.; William Dam Seeds; Glecklers Seedmen; Grace's Gardens; J.L. Hudson, Seedsman; Kitazawa Seed Co.; Lakeland Nurseries; Earl May Seed and Nursery Co.; Mellinger's, Inc.; Nichols Garden Nursery; Geo. W. Park Seed Co.; and R.H. Shumway Seedsman.

Fava Bean

Vicia faba

THE fava was the only edible bean in Europe until explorers made it back from the Americas with kidney, pinto, and lima beans. Remnants of this ancient food were found in Swiss lake dwellings from the Bronze Age, and the ancient Egyptians, Greeks, and Romans grew it. In 2822 B.C. the favored fava also was being eaten in China. Today this attractive and soil-improving relative of vetch is a popular crop in Europe, the Middle East, Egypt, India, and Burma, as well as in Mexico and Brazil, where it's sown at higher altitudes. These widespread growing sites have resulted in many a "nom de bloom" for this vegetable. Known in southern Europe as the faba and in California as the fava, the brisk-climate bean is called the broad bean in Britain and Canada and sometimes the shell bean, English dwarf bean, or the horse bean. As that last name unfortunately suggests, in the United States the fava is still grown mostly for fodder or as a winter cover crop that's turned under. This neglect by home gardeners is really too bad, for the fava's flavor is a pleasant blend of lima and pea, and it thrives in the cool rainy locales that kill off snap beans.

The fava is an upright annual that sends forth succulent-looking blue green leaves from its squarish and pulpy stalks. Somewhat delicate, fava stems are without tendrils and start to need support when they reach 2 feet or so. The plant spawns handsome white to pink blossoms with distinctive black centers. These are followed by almost-round green pods. In the so-called WINDSOR types of broad beans, these pods are short and each contains about four large, flat seeds that resemble limas. LONG POD favas, however, have pods up to 12 inches long, each with about eight smaller, oblong beans.

Needing a long, cool growing season of 90 days, favas have roughly the same climatic requirements as peas. They are widely grown in England and in eastern Canada, for though they can't survive heavy ground freezes, they handle frost with aplomb, and often stands wilted and dulled by nippy nights will make a dramatic return to life if the weather warms. (It helps if frosted plants are sprinkled before the sun gets to them.) Despite the broad bean's hardiness, much of the United States has winters so cold that fall plantings are out of the question. And summers in the South are too hot—for if the air temperature goes above 70°F. before the plants set pods, the blossoms will drop off without opening. But in the Gulf Coast states and lower California, the nitrogen-fixing bean makes a good winter crop, succeeding snap beans and preceding corn, squash, and other hot weather vegetables. Planted from October to December in such mild climates, it will be ready for harvest in March to May.

In Algeria, the fava is still found wild, but there are many improved varieties on the market. Most strains grow over 3 feet high, but you also can buy dwarf varieties like the SUTTON and the MIDGET, which don't need support. STAYGREEN is a semi-dwarf, and AQUADULCE CLAUDIA is recommended for autumn planting.

Sow your broad beans as early as possible in the spring—even if your soil is still a bit icy. Place them 2 inches deep in rows 18 to 20 inches apart, sowing about five to eight seeds per foot. (Don't, by the way, bother to treat your favas with ordinary legume inoculant, for they need a special kind developed for vetches.) Thin the seedlings to stand 3 to 4 inches apart. Like other legumes, favas are tolerant of pretty poor soil, but they prefer well-cultivated ground that has been manured a few months earlier.

If you're in a warm region, your plants may bear better and longer if you arrange matters so they get shade for part of the day. As the season wears on, you also should hill soil up around your plants and use mulch to help keep them cool. You can support the fragile stalks with a minimum of trouble by running old clothesline around stakes placed at both ends of each row. Or plant in a bed near a fence so you can encircle your stand with a rope anchored to the fence if high winds should come.

Favas sown in spring may fall prey to certain insect pests with the coming of hot weather. To combat black aphids, pinch out the tops in full flower and burn the affected parts. Pick off any bean beetles and green stink bugs, using rotenone if there are too many bugs to remove by hand. Given the vulnerability of favas when the temperatures climb, you may find it worthwhile to start your seeds in the greenhouse or in a cold frame in December or January so you can set plants out in March and harvest them before the heat sets in.

The pods nearest the soil will be ready for gathering first, so you can harvest upward at five-day intervals. If you pick the pods when they're from 2 to 3 inches long—before the cottony texture develops on the inside—you can use them like snap beans. Or you can shell the small seeds, which are much like peas at this stage, and cook them as you would peas. If you harvest and shell the beans when they are older, it's best to remove the outer, light green part and stew or simmer only the inner dark green pea-like portion. Assuming your patch is untroubled by insects, you might want to let the pods mature and the beans dry on the plant. In wet weather, though, you'd best finish the drying job inside so the beans don't rot in the pod. You also can quick-freeze or can young broad beans in the pod or the shelled mature beans.

Sweeter than limas, favas make a welcome dish in early spring when something green on the ground and fresh in the mouth is much appreciated. The immature vegetable offers as much protein as green limas, ten times as much as snap beans, and about a quarter more than peas—plus more iron and potassium than any of these. And dried broad beans have about half as much protein as dried soybeans.

Young favas in the pod are delicious simmered, steamed, or stir-fried, while shelled older beans taste good in hearty dishes with ham, pork, or bacon. The French savor them pureed or with heavy cream as an accompaniment to meat. You might also boil good-sized favas in the pod, then cool them and eat from the pod the way southerners enjoy boiled peanuts. For good snacking, pop dried favas like popcorn and eat with salt as the Japanese do—or roast them like peanuts. You can also salt-pickle the dried beans as they do in the Middle East. Try the young leaves of the plant too—they're edible, and make a tasty potherb.

Order your fava beans for planting from Burgess Seed and Plant Co.; W. Atlee Burpee Co.; Comstock, Ferre and Co.; William Dam Seeds; J.A. Demonchaux Co.; DeGiorgi Co., Inc.; Le Jardin du Gourmet; Joseph Harris Co., Inc.; Charles C. Hart Seed Co.; Johnny's Selected Seeds; Nichols Garden Nursery; L.L. Olds Seed Co.; Seedway; R.H. Shumway Seedsman; Stokes Seeds, Inc.; Thompson and Morgan, Inc.; or The Vermont Bean Seed Co.

A FAVA BEAN FLAW

Though the broad bean is cultivated and used as a food in almost every country in the world, some few individuals—usually males of Mediterranean descent—experience a hereditary allergic reaction when they inhale fava pollen or eat its seeds. The symptoms of favism, as it's called, usually disappear without treatment within a few days of exposure. This problem is usually linked to fresh (rather than dried) beans, and the highest recorded incidence of it recently occurred in Sardinia, affecting only 5 people per 1,000 population.

Since the fava is a member of the vetch family, its seeds also may contain a toxin capable of creating muscular weakness and paralysis when the seeds make up all or most of a person's diet.

The substance causing this lathyrism is linked only to certain species—or even varieties—of vetch and may not be present in the seeds under all conditions. In any case, the disease is associated only with tremendous consumption of the food involved, and eaten in moderation in a balanced diet, broad beans don't seem to produce an ill effects. As an extra precaution you might want to soak or cook the beans in hot water, which you then discard. You'll lose some water-soluble vitamins along with any toxin that might be there, but you'll still come through with the fava's excellent protein (compared to the ideal protein scoring 100, favas rate 67, versus 68 for soybeans, 55 for kidney beans, and 52 for peanuts).

Horticultural Bean

Phaseolus vulgaris

THESE versatile beans, sometimes referred to as "shell beans" or "wren's egg beans," deserve to be better known than they are. Horticultural beans are eaten primarily as green shell beans, but they also make tasty snap beans if eaten before they mature, and can be dried, too. The slender, rounded pods of the horticultural bean usually grow to be about 5 inches long, although some varieties get as long as 8 to 9 inches. The pods are light green in color when young, and develop a pretty pink or red mottling as the beans mature. The beans inside, when mature, are pink and buff colored or, occasionally, a solid raspberry pink.

Horticultural beans come in both pole and dwarf bush varieties. They both do well in cooler climates, and are grown like regular pole or bush green beans. If you are planting horticultural beans along with your standard bean crop, be sure to mark the rows in which you plant them. When they are young and green, horticultural beans look just like snap beans, and you could quite easily harvest the whole crop by mistake and miss out on the mature beans.

Your crop will thrive on a rich, humusy soil that is kept well watered. Keep a close watch throughout the season for threats of an aphid invasion. If one seems likely, give the beans a good hosing down with a heavy spray of water to get rid of the pests.

• When I was growing up near Boston, the green-grocers' markets in August were always well stocked with what we called "Shell Beans.". . . Then for years these beans seemed to disappear from my life, and it wasn't until after I was married and my husband and I grew our own vegetables that I discovered these shell beans were to be ordered from the seedsman under the name "French Horticultural Beans." We've grown a packet or so of this kind of bean ever since. . .

One year I invented another way to prepare these beans in a recipe which turned out to be very popular with my family and friends. All you do is simmer the fresh beans until tender but not mushy. Then drain them and put them into a lightly seasoned French dressing of 3 tablespoons of vegetable oil, 1 tablespoon of vinegar, and a little salt and pepper. To this you add several cloves of garlic, chopped fine. Let them sit and cool in this mixture for several hours (or more) and serve them speared on stout toothpicks (I put on about three) or as a side-dish salad. Anyone who likes garlic finds this dish irresistible.

—*Catharine O. Foster*
Bennington, Vermont

31

If your garden is bothered by Mexican bean beetles, it would be wise to plant the beans near the potatoes to help control the beetles.

Pick the pods before the beans inside develop if you want to use them as snap beans. But by all means let some of the beans mature—horticultural beans are at their best when eaten as green shell beans, fresh from the vine and served piping hot with a touch of butter. For green shell beans, pick the pods as soon as they fill out. The crop usually matures pretty much all at the same time, so be prepared to can or freeze some of the last couple of pickings. To can, pack the shelled and washed beans loosely in jars to within 1 inch of the top for pints and 1½ inches for quarts. Process in a pressure canner at 10 pounds pressure, 50 minutes. To freeze, wash the beans, blanch for 2 to 3 minutes, cool, pack, and freeze. You might also like to dry some of the beans for use in your favorite dried-bean dishes. In that case, let the beans mature fully on the vines, then cut the plants and allow them to dry thoroughly in an airy place. When the beans are completely dry, they are ready to be shelled and stored. To guard against insect invasion during storage, freeze the beans for a few hours before storing, or heat them in a slow oven (250°F.) for 10 to 15 minutes. Then store them in jars or other tightly covered containers, just as you store other kinds of dried beans.

After the entire crop has been harvested, turn the plants back into the soil to make the most of their nitrogen-fixing bacteria. If, for some reason, you decide not to bury the bean plants in the garden, they will prove valuable in the compost heap.

Immature horticultural beans can be prepared in the same ways you prepare regular snap beans. Try them steamed and topped with butter and your favorite herbs, or cheese; or stir-fry them with garlic and onion. The delicious green shell beans can be steamed or cooked in a small amount of water until tender, and served any way you serve fresh lima beans. They are superb served with a garlic cream sauce, or topped with sour cream and dill. For a colorful variation on an old favorite, combine some cooked horticultural beans with leftover cooked sweet corn you've shaved from the cob, add a bit of butter, salt and pepper and a tablespoon or two of cream, heat through, and you've got a luscious succotash. Mature dried beans can be used in all your favorite dried bean recipes, such as baked beans and bean soup.

Horticultural beans are available from several companies, including W. Atlee Burpee Co.; Comstock, Ferre and Co.; Gurney Seed and Nursery Co.; Joseph Harris Co., Inc.; H.G. Hastings Co.; Charles C. Hart Seed Co.; Johnny's Selected Seeds; Earl May Seed and Nursery Co.; L.L. Olds Seed Co.; Seedway; R.H. Shumway Seedsman; Stokes Seeds, Inc.; Otis S. Twilley Seed Co.; and the Vermont Bean Seed Co.

Purple Bush Bean

Phaseolus vulgaris

LISTED in most seed catalogs as a novelty bush bean, such a classification belies the usefulness of this bean. Instead of being a mere novelty, furnishing a conversation piece for the gardener, purple beans are beautiful in the garden, appetizing on the plate, and delicious when eaten. They share hardiness and the ability to ward off insect pests with the other purple vegetables (in particular, purple cauliflower). They also seem to be more tolerant of cool, wet weather than other bush beans. Such value cannot be put down as passing novelty.

Purple bush beans are sold under the names of ROYALTY or ROYAL BURGUNDY. They are as prolific as most bush beans, if given the proper care. A California gardener reported that he once made the mistake of growing purple beans on soil that was not well conditioned for gardening (the ground was new to him and his efforts), and he found that they produced short, misshapen pods with only one or two beans per pod. Another gardener from North Carolina described a similar problem. The moral of the story, then, is to treat your purple beans with the same respect you would give to any bean, planting them in full sun and in well-prepared beds. They do best in ordinary soil that has a pH between 5.5 and 6.7, and is fairly low in nitrogen.

Plant the seeds over an inch deep in rows 10 to 12 inches

apart as soon as the soil has been well warmed by the spring sun and there is no chance of a surprise frost. If you space the seeds about 4 inches apart, no thinning will be necessary later. The few beans which don't germinate leave only small gaps which can be later filled in with summer savory plants. The savory is an insect-repellent herb which makes a good companion to beans, both in the garden and in the kitchen.

Twenty feet of row should be more than adequate for a family of four who plan to freeze some of their produce. Purple beans produce larger, more vigorous plants than other bush beans—they are more viney than bushy. This may account for their tendency to produce beans over a longer season than most green or yellow varieties. The beans are harvested when the purple still has a velvety surface, and before the grain of the bean begins to develop. The seed matures rather slowly in the pod, so the beans remain edible for several days after they reach picking size. This is an asset to beleaguered gardeners who are hard-pressed to keep up with the midsummer harvest. Like conventional kinds of green beans, the sooner you use purple beans after picking them, the crisper and tastier they will be.

Many people consider the purple bean the best-tasting bush bean available, describing the flavor as tasting like green beans, but better. It is also said to be the best variety for freezing. When the beans are cooked, their purple hue changes to bright green (an unending source of entertainment for kids). This color change acts as a built-in timer for the gardener who wants to freeze these beans. When you see the green, you'll know you have done just the right amount of blanching to freeze them properly. Purple beans will take to canning as readily as other snap beans. All you need do is wash the beans, trim the ends, and break into 1-inch pieces. Pack the beans tightly in jars, to within ½ inch of the top. Process at 10 pounds pressure, 20 minutes for pints and 25 minutes for quarts.

You can use these beautiful beans in all your favorite snap bean recipes. They're terrific with mustard sauce, au gratin, with mushrooms or shallots, and served sweet and sour. If you want to save the seed from year to year, this bean, like all others, will give you the true bean at next planting.

Purple beans are available from many seed houses, including Burgess Seed and Plant Co.; W. Atlee Burpee Co.; Farmer Seed and Nursery Co.; Gurney Seed and Nursery Co.; Johnny's Selected Seeds; Lakeland Nurseries; Earl May Seed and Nursery Co.; Nichols Garden Nursery; L.L. Olds Seed Co.; Geo. W. Park Co.; R.H. Shumway Seedsman; Stokes Seeds, Inc.; Thompson and Morgan, Inc.; and The Vermont Bean Seed Co.

Purple Bush Bean

Romano Bean

Phaseolus vulgaris

FOR every gardener who insists that the purple bean is
the king of the snap beans, there is another equally adamant
gardener who steadfastly maintains that the honor belongs to
the Romano bean alone. The Italian pole bean, as it's also
called, has found its way into many a garden across the United
States, and once gardeners have tried it, many of them never
grow any other green bean. What's the attraction of this rather
ordinary-looking bean? First and foremost, the flavor. When
not overcooked, the Romano bean possesses a delicate and
complex taste all its own which makes it a natural accompaniment to Italian and southern European dishes. It is welcomed
by vegetarians looking for a new taste in beans, as well as less
sophisticated bean eaters, who find that it tastes "like green
beans but different." And it's a good crop for gardeners
without a lot of room, because the plant's climbing habit allows
them to get more production out of smaller spaces. In more
and more gardens, the standard varieties of string beans are
partially or even completely giving way to the prolific Romano
bean. A bush variety of this popular bean, the ROMA bean,
has been developed as well, for gardeners not fond of growing
pole beans.

Italian pole beans grow on medium-sized vines that can be
trained up a trellis, fence or poles to save ground space.

Horizontals are of no use to them, but given plenty of vertical support, these climbers will go to town. If you are growing them in a greenhouse, the vines can be trained right up the sides of the house to save bench space. The vines grow tall and should be confined to the north wall so that they will still let in plenty of light for your other plants.

Like all beans, Romanos should not be planted until the soil is really warm and all chances of frost are past. Plant the seed 1 inch deep and 4 inches apart in loose, rich soil, after 7-foot poles have been set up for the plants to climb. The beans will sprout in about 10 days, and the bed should not be kept overly damp during this period. Once established, the plants will generally tolerate a wide range of growing conditions. An 8-foot row will provide plenty of beans for a family of four. A 20-foot row will provide the same family with ample beans for freezing and drying.

Like all pole beans, Romanos flower from the bottom up. By pinching out the bottom-most flowers you can avoid having beans that touch the ground and need to have the dirt washed from them. This pinching back also seems to stimulate the growth of new flowers.

If bean beetles are a problem where you live, planting your Romano beans near potatoes, nasturtiums, or summer savory should help. Another good trick is to plant a clove of garlic by each pole. These measures will probably be sufficient to drive away any invaders, as Italian pole beans are extremely hardy and resistant to pest attack.

Romano beans are most often used in their immature stage, as snap beans, but they can also be allowed to develop fully to be dried for winter use. For use as snap beans, Romanos should be picked while they are still crisp, before much cellulose builds up in the wall of the bean. The broad, flat pods will still snap when quite large, and any strings that have developed can be easily removed. Like all beans, keep them picked to keep them coming.

Immature Romano beans cook quickly—more quickly than ordinary green beans. Exercising care not to overcook them will prove well worthwhile, for the marvelous flavor of these beans

• My friend Able Ferreira took me by the arm to his garden to show me his "Portagee Beans." He forced a bucket of them on me, and I took them home not expecting much from these large, crude-looking beans. I was pleasantly surprised to find that the beans were delicate, they cooked faster than ordinary green beans, and they had a complicated, very compelling taste. I was sold.

On investigating I found that the Portagee beans were actually an old bean of the Mediterranean, used by the Italians, Spanish, and French alike. In this country it is known as the Romano bean.

Try it cold in vinaigrette sauce. Try it as a side dish to spaghetti. Try it as beans with pesto sauce. Try it, but don't overcook!

—*John Meeker*
Gilroy, California

when properly cooked defies description. Romano beans need only the subtlest of seasonings or the very lightest of sauces to show off their best qualities. They are delectable when served with a light mushroom or onion sauce. To freeze them, you need parboil them only slightly. They may be canned, too, by the same method given for purple beans.

At the end of summer, all the newly formed beans at the top of the vine can be allowed to mature and dry. In a small row, this top growth at the end of the season will give you over three pints of large dried beans. Try them in your favorite bean soup recipe for a hearty mid-winter treat.

Romano pole beans or Roma bush beans are available from many sources: Burgess Seed and Plant Co.; W. Atlee Burpee Co.; Comstock, Ferre and Co.; William Dam Seeds; DeGiorgi Co.; Gurney Seed and Nursery Co.; Joseph E. Harris Co., Inc.; Charles C. Hart Seed Co.; H.G. Hastings Co.; Jackson and Perkins Co.; Earl May Seed and Nursery Co.; Nichols Garden Nursery; L.L. Olds Seed Co.; Geo. W. Park Seed Co.; Redwood City Seed Co.; Seedway; R.H. Shumway Seedsman; Stokes Seeds, Inc.; Thompson and Morgan, Inc.; Otis S. Twilley Seed Co.; and The Vermont Bean Seed Co. are some of them.

Romano Bean

Scarlet Runner Bean

Phaseolus coccineus

BELIEVED to be native to South or Central America, the Scarlet Runner has been grown in the United States for many years, primarily for its brilliant red blossoms. The runner bean, also called the Multiflora or Painted Lady Bean, was introduced to Europe during the seventeenth century, but was regarded there strictly as an ornamental until nearly a hundred years later. Today the Scarlet Runner is the most popular green bean in England, and is highly regarded in continental Europe as well.

In its native climate the runner bean is a perennial whose bulbous root lies dormant in the ground during the winter, sending up new shoots the following spring. The root may survive a mild winter further north, but in this country it must generally be treated as an annual. That means either planting new seeds every spring, or digging up the tubers in fall and storing them, like flower bulbs, indoors for the winter.

The runner bean is a vigorous climber, often ascending 10 or 12 feet in a season. The plants can be trained up poles in the garden; the growing tips are pinched off when the plants reach the top of the poles. The plants will also climb up trellises, trees, or strings run up the side of the house. The bright scarlet flowers held well out from the profusion of dark green leaves form a lovely ornamental screen, while serving the dual purpose of producing edible beans later on. Runner beans can be eaten as snap beans when the pods are young, in their imma-

In A Nutshell. . .

I would like to recommend the Scarlet Runner bean . . . This plant has many admirable qualities. It makes a beautiful privacy screen with its lush green foliage; it grows fast and produces striking scarlet blossoms. . . Soon, edible green beans appear in great numbers. These may be harvested from July through September.

The scarlet blossoms attract great numbers of bees and hummingbirds, but the plants are not bothered by insects.

—*Janet Studer*
East Syracuse, New York

43

ture form as green shell beans, or matured for dried shell beans.

The flowers of the runner bean are usually red (hence the name), but there is also a white-flowered Dutch variety. The broad, flat pods grow 8 or 9 inches long, and develop large seeds which vary in color from white (from the white-flowered plant), through assorted shades of pink, sometimes mottled with purple or black. Observant gardeners will notice two peculiar things about this plant. First, unlike other common beans, its cotyledons (first leaves) germinate underground; second, the plant twines from right to left, whereas most plants twine from left to right. These exceptional characteristics aside, the Scarlet Runner is cultivated much like other pole beans.

Plant the beans in spring, after the soil has warmed to 55 or 60°F. A crop of young pods can be had in just about any climatic zone, but a crop of mature seeds requires 115 to 120 days. If your growing season is short, you'll have to start the seed indoors in order to harvest dry beans before frost.

Although they're not really fussy, runner beans do appreciate a rich, light, well-cultivated soil that's not too acid. Preparing the soil with compost or manure is sound practice, although you want to avoid an excess of nitrogen. If your compost is quite high in nitrogen, it's a good idea to balance it with some potash-containing material. Sow the seed in rows, 1 to 2 inches deep and 6 inches apart. When established, the seedlings should be thinned to stand a foot apart. The plants can also be trained up poles fastened together at the top to form tepees; in this case, sow seed at the base of each pole, and set the teepees a foot apart.

A World Record?

In summer of 1976, the Vermont Bean Seed Company reported what they believe to be the longest Scarlet Runner bean pod in the world—14½ inches. The plant it came from was 100 percent organically grown, grew 14 feet high, and kept right on producing through two frosts.

Mulch the plants well to hold in moisture. During dry weather they may need to be watered as well. The flowers are pollinated by bees, and planting borage nearby will help attract lots of them.

Scarlet Runners are not generally bothered by insects. If you have slugs in your garden, though, they are likely to attack the young plants. Some strategically placed shallow dishes of beer should keep the slugs happy, and away from your beans.

Keeping up with the runner bean harvest encourages continued production. When harvesting Scarlet Runners for use as snap beans, pick the pods while they are still young and tender. Older pods become tough and stringy. The pods are at their best if French-cut in lengthwise strips before cooking. They can then be prepared just like ordinary snap beans. Try them baked in mushroom sauce, sauteed with garlic and topped with herbs and grated Parmesan or in your favorite snap bean salad. At this stage, you can can or freeze the beans as you do snap beans. The green shell beans, picked as soon as the pods fill out, are prepared like fresh lima beans; they're quite tasty with herb butter, sauteed shallots, or cheese. Can or freeze them like lima beans (the methods are described in the section on horticultural beans). Fully mature beans are dried and used like other dried beans. Scarlet Runner beans have a pleasant, rather nutty flavor when cooked.

Suppliers of Scarlet Runner beans include: Comstock, Ferre and Co.; William Dam Seeds; Henry Field Seed and Nursery Co.; Joseph Harris Co., Inc.; J.L. Hudson, Seedsman (under the name of Painted Lady Bean); L.L. Olds Seed Co.; Redwood City Seed Co.; R.H. Shumway Seedsman; Stokes Seeds, Inc.; and the Vermont Bean Seed Co.

TANGY BEAN SALAD

1 pound fresh Scarlet Runner beans
½ medium red onion, thinly sliced
1 clove garlic, minced
6 tablespoons olive oil
2 tablespoons red wine vinegar or tarragon vinegar
½ teaspoon salt
　freshly ground pepper to taste
　pinch paprika
1 teaspoon prepared mustard or a pinch of dry mustard

Wash the beans, snap off the ends, and French-cut into lengthwise strips. Steam briefly, until just tender; drain thoroughly, and place in a large bowl with the onion and garlic. Toss well. Combine the remaining ingredients to make a dressing. Pour over the beans, toss well, and refrigerate for an hour or two before serving to blend flavors. Serves 4 or 5.

Soybean

Glycine max

SOYBEANS grown for drying and fermenting into flavorful foods have nourished the Orient for over 5,000 years, and the varieties harvested as a green vegetable are at least a venerable 1500 years of age. Traditionally considered one of the five sacred foods essential to Chinese civilization, soybeans were being grown in the botanical gardens of Europe by the mid-eighteenth century, when Benjamin Franklin sent some home to America from France. In 1850 more beans reached Illinois and Ohio, coming from Japan via San Francisco. Commodore Perry also brought soybeans back from Japan in 1854, but by the end of the century only a few varieties were known in the United States—and none were vegetable beans.

Around that time, however, the U.S. Department of Agriculture became interested in this drought-resistant plant and imported an amazing 10,000 varieties from China for testing. A handful of these introduced types proved well-adapted to the prairie states, and today the United States ranks with China as the leading grower of field soybeans, which are now the most economically important bean in the world. Generally, though, the larger, more tender strains savored as green "branchbeans" in Japan and China have continued to be overlooked in this country. In fact, it's only in the last 30 years that these green or vegetable soybeans have become known here. Too

47

bad, too, because their high protein content in combination with a deliciously fresh flavor and nitrogen-fixing ability make them a natural for home gardens.

A branching plant that can be either erect or somewhat prostrate in habit, the soybean looks like the common or navy bean in its early growth. Its trifoliate leaves vary widely in shape and size, and so do the plants, which range from less than one foot to several feet in height at maturity. Stems, leaves, and pods are usually covered with grey or brownish fuzz that seems to offer some defense against insects. Reaching deep into the ground, soybeans have a more or less pronounced taproot that makes them fairly tolerant of drought. Adapted to about the same climates as corn, like other legumes soybeans have roots hosting nodule-forming bacteria that can convert the nitrogen in air to a plant-usable kind. Soybean flowers are small, hard to see, and self-fertilizing, and the rate at which the plants flower and produce pods with one to four beans depends on day length. This means that varieties bred in the South usually don't mature before frost in the North, whereas those adapted to the North will flower and set seed very early in the short days of the southern states.

Though field soybeans need a very long growing season in order to ripen and dry in the garden before harvest, growing varieties you plan to harvest as green shell beans makes the slow-germinating and slow-to-set-seed soybean a viable crop even if you live in a climate with short, somewhat cool summers. (Be wary, though, if your summers are damp and overcast as in the Pacific Northwest, since even the recently developed short-season types can't do well under such conditions.)

In choosing from among the many varieties that taste good green, think first about your growing season, for there are soybeans that require well over 100 days just to reach the green stage in northern areas. The widely offered KANRICH is one such bean, and like the late varieties SEMINOLE and ROKU-SAN, it's best suited to the South and to gardens as far north as southern Pennsylvania, where it is green-bean ripe in about 82 days. The same is true of JOGUN and of the high-yielding HOKKAIDO, which is table-ready at about 80 days in southeastern Pennsylvania. FUJI and BANSEI (which freezes well) are ready for harvest somewhat earlier. So are EXTRA EARLY GREEN and EARLY GREEN BUSH. If you happen to garden in a really brisk climate, you can benefit from the 40 years a Swede named Sven Holmberg spent developing an edible soybean that will crop as far north as southern Canada—a bean that actually yields much less when grown in the longer, hotter summers of "southern" states like Pennsylvania. The FISKEBY V VEGETABLE BEAN, as it's called, is a rather small but highly ornamental plant that can be used in decorative land-scaping. Producing mostly three beans per pod, it takes a remarkably short 68 days to mature and contains a whopping 39 to 40 percent protein. The Fiskeby is reportedly very low in the antitrypsin factor that interferes with the digestion of uncooked soy protein, which means you can enjoy this variety out of hand in the garden or in salads.

Another good bet for short-season gardeners is the high-yielding PICKETT, which is a full-sized plant with large beans ready to eat at 85 days. One New England gardener deter-mined to do the impossible—namely, grow soybeans to matur-ity near Burlington, Vermont—got excellent results with OKUHARA, AKITA, ALTONA, and ENVY (a variety de-veloped for the Durham, New Hampshire, locale). All were harvestable at about 90 days, with Altona yielding three-seeded pods of very big beans and Akita coming forth with the largest beans and healthiest-looking crop of all, with one or two beans to the pod. If you have warm to hot summers on the dry side, you might like to try the very small BLACK SOYBEAN. These heavy producers are sweeter than any other bean when cooked and star as a high-protein holiday confection in Japan. The pods of these black soybeans can be left on the plant to dry, for unlike many varieties, they don't shatter as the foliage begins to yellow. (By the way, if you're planning to let some of your soys dry in the garden, it will be well worth your while to shop around for a shatter-resistant variety like Kanrich.)

Start vegetable soybeans in late spring, when all danger of frost is past. Sow at 4-inch intervals—less if you're growing a low-growing variety—and about 1½ inches deep in well-prepared soil. Soybeans like a sunny location and light, warm, loamy soil. It's best to fertilize only lightly since the beans will take longer to mature in rich soil. Besides, it's known that while fertilizer on poor soil can increase yield markedly, this does not happen when already-rich soil is treated further. To boost your germination rate, speed up early growth, and increase yields, you might want to moisten the seeds and shake them immediately before planting in a bag containing legume inoculant. Available very inexpensively from many seedhouses, these organic bacteria in powder form will help your crop fix nitrogen.

If you're after high yields, light cultivation around the soybean plants is a good idea, for these slow-growing youngsters can be outstripped by weeds rather quickly. Disease and insects usually don't pose much of a problem, but cool rainy weather favors bacterial blight, which will show as small angular brown to black spots. Downy mildew, on the other hand, can create small pale green spots on the leaf tops, and bacterial pustule causes spots with reddish brown centers. All these diseases can be carried over the winter on diseased stems and leaves, so if you see any sign of them, destroy the infected plants, plow deeply, and rotate soybeans with another crop the following year. Be sure also that your seed is disease-free. If Japanese beetles are a pest in your area and begin munching on your plants in August, you might want to buy spore-powder for the biological control known as milky spore disease. This bacterial organism is quite safe to use and destroys beetle grubs when added to infested soils. Another potential pest in the Southeast is the fast-chewing velvetbean caterpillar, which lays small white eggs on the surfaces of the leaves. You can pluck off the eggs or caterpillars by hand, but if matters reach epidemic proportions, try the safe biological control Thuricide (*bacillus thuringiensis*).

Vegetable soybeans can be grown in a cold frame if necessary and do well in a greenhouse when you sow three seeds in a 10-inch clay pot. If you prefer to plant them in the bench, sow the beans 1 inch apart and ¼ inch deep and train on a wire fence or trellis.

Pick your edible soys when they're fully formed in the pods but still green. This will be before the foliage begins to yellow. Any excess will freeze beautifully. Just wash and blanch your unshelled beans for 5 minutes in steam or boiling water. After you cool and shell the beans, rinse them in cold water, then pack and freeze. To can your harvest, shell the beans, then cover with boiling water and bring to a boil. Drain, reserving liquid, then pack the beans loosely in a jar and cover with hot liquid, leaving 1 inch at the top. Process in the pressure cooker at 10 pounds pressure—55 minutes for glass pints and 65 minutes for glass quarts. For much simpler long-term storage, you can sacrifice the fresh lima-like taste of some of your beans by letting them dry on the plant. Pick the newly dried beans while the stems are still green, though, or the shells will open and scatter your protein to the four winds. If these dried beans are harvested late enough at the end of a fairly dry season, they'll pop out of their shells easily when you put them in a cloth or paper bag and swat them with a heavy stick.

Though your soybean yield may not look huge, these smooth buttery-tasting beans have a high concentration of protein, with over 16 grams in a cup. That's more than any other green vegetable. It's high quality protein, too, for soy protein has one of the best balances of amino acids in the vegetable kingdom, and when combined in dishes with whole wheat, rice, milk, and cornmeal, soybeans can help create protein as valuable as that in meat. Though they don't have as great a concentration of protein as dried soybeans, the green ones do offer much more vitamin A and vitamin C than you'll find in the same amount of dried soys.

Try some of your beans-in-the-pod as *edamame,* or branch-beans, a Japanese-style summer snack. Put the washed beans in boiling salted water, cook until tender, then drain and chill. Serve these along with some cold beer or iced tea and watch your guests have fun popping the cool beans into their mouths straight from the pod. Shelled beans also make a delicious steamed vegetable served with butter or can be stir-fried in a little oil alone or with other vegetables. Try them, for instance, with tomatoes, green onions, and green peppers.

For a scrumptious salad, marinate shelled and lightly cooked beans in oil and vinegar, adding chopped onion, celery, shredded carrot, green pepper, fresh dill, and salt and pepper. Or try them with a seasoned cream sauce, or baked in tomato sauce.

If you choose an early variety and intercrop it with high-lysine corn that matures at the same time, the delicious succotash you can serve will provide ample complete protein, making a meat entree unnecessary. You can create other low-cost complete protein combinations by serving vegetable soybeans in a casserole or soup with homemade whole wheat bread on the side, or stir-fried with rice and soy sauce oriental-style, or as part of a Mexican meal featuring cornmeal.

For more fun, roast and grind some of your dried beans and use them like coffee. After all, "coffee bean" and "coffee berry" are old names given to the soybean when Americans were still using it mostly as a forage crop or for hay or silage. Happily, we now know better, and you can buy green soybeans for your garden from many seedsmen. Sources include Burgess Seed and Plant Co.; W. Atlee Burpee Co. (Kanrich); Comstock, Ferre and Co.; William Dam Seeds; DeGiorgi Co.; Gurney Seed and Nursery Co. (Disoy); Johnny's Selected Seeds (Envy, Altona, Traverse, Meredith, Giant Green); Kitazawa Seed Co. (Extra Early Green); Le Jardin du Gourmet; Earl May Seed and Nursery Co.; Nichols Garden Nursery; Stokes Seeds, Inc. (Early Green Bush, Verde); Thompson and Morgan, Inc.; (Fiskeby V, Okuhara); and the Vermont Bean Seed Co. (Black Soybean).

Borage

Borago officinalis

BORAGE is believed to have originated in the Middle East, and has been well-known throughout the world since ancient times. The Romans made its flowers and leaves into an elixir which Pliny said had the power to make men joyful, and their hearts merry. Borage was also popular in Elizabethan England. This unassuming herb, with its star-shaped, sky-blue flowers, was often prescribed in those days for melancholy, and it has over the years come to be associated with the idea of courage. A garden without borage, it is said, is like a heart without courage.

In the colonial era of our own country, many gardeners grew borage, and it is still found in herb gardens today. Borage is an easy-to-grow annual that will reseed itself once established. Keep this in mind when you plant your first crop and give it plenty of room to expand—about 2 feet in every direction. The plants grow quickly, and will soon bear huge numbers of beautiful blue flowers. Their high nectar content and glorious profusion make them irresistible to bees. Occasionally a single plant may produce some pink flowers along with the masses of blue ones. Don't be surprised; this mutable capacity is characteristic of borage.

Except for the flowers, mature borage is not the prettiest of plants. The leaves and stems are coarse and ungainly looking. Still, planting it in a close-spaced bed will give you a dense,

Borage Tea for a
Multitude of Ills

Lots of folks enjoy iced borage tea as a summertime beverage. Take a handful of fresh or dried borage leaves, steep in water that has been brought to a boil, chill, add ice, and serve with borage flowers floating on top. The same drink, served as a hot tea, has an ancient reputation as an invigorating tonic, and is considered especially good for the lungs. Borage tea is listed in many old herbals as a demulcent, emollient, and diuretic, and a drink made with borage and lemon was prescribed to reduce fever.

bright stand reminiscent of a tropical jungle. You won't even notice the clumsiness of the plants.

Borage responds readily to a light, rich soil fortified with composted manure. Catalogs usually list borage as growing about 2 feet tall, but some gardeners have watched their plants, in rich soil, branch farther and farther out to reach a height of 4 feet. Cultivate the bed regularly to keep the earth loose, well-aerated, and free of weeds. A layer of mulch will help maintain the kind of moisture-rich environment borage likes so well.

If you are planting borage strictly as a vegetable crop, you might prefer to plant it in rows rather than beds, for easier accessibility. In this case, the borage can be planted just as you'd plant spinach.

Borage will also do well in an indoor garden or greenhouse. It's quite content in a pot, so long as its roots have enough room, the potting mixture is fertile, and moisture and daylight are in ample supply. Its cheerful flowers will lend a note of brightness to a winter-weary household.

Borage flowers are sometimes candied and used for decorating cakes and other confections. To make your own, dip the flowers in a bit of egg white, then in sugar. It's fun to use the flowers, but it's the plant's leaves that are truly important. They possess a mild, cucumber-like flavor guaranteed to perk up any salad. Harvest the leaves for eating by pinching them off when they are 3 to 5 inches long. Larger leaves are not as palatable. The older plants become "leggy," and develop tiny, hairy spines. The spines aren't sharp "stickers," but they do make eating fresh or stir-fried leaves an irritating proposition. Finely chopping the leaves might help, but it's better to eat only the young ones. If you want a constant supply of fresh young leaves, sow a new row of borage every few weeks throughout the season.

Even though the flavor of borage is most often compared to cucumbers, don't be fooled into using the herb as a cucumber substitute. The taste of borage is so distinctive, you will either think it is one of the most wonderful vegetables in existence, or else consider it a noxious weed unfit for the

human palate. But don't let this all-or-nothing declaration keep
you from growing borage. Try the young leaves in tossed green
salads, or cooked and mixed with spinach or mustard greens
and chopped hard-boiled eggs. A bit of borage added to a glass
of lemonade makes for an extra-refreshing thirst quencher. If
you decide the taste of borage is not for you, just stop picking
and wait for the lovely blossoms to appear. Or share the extra
greens with your chickens. They love borage, and you'll benefit
with healthier chickens and tasty eggs.

Borage is available from Casa Yerba; Gurney Seed and
Nursery Co.; Joseph Harris Co., Inc.; Charles C. Hart Seed
Co.; H.G. Hastings Co.; J.L. Hudson, Seedsman; Le Jardin du
Gourmet; Johnny's Selected Seeds; Meadowbrook Herb Gar-
den; Mellinger's, Inc.; Nichols Garden Nursery; L.L. Olds Seed
Co.; Geo. W. Park Seed Co.; Redwood City Seed Co.; R.H.
Shumway Seedsman; Stokes Seeds, Inc.; Otis S. Twilley Seed
Co.; and Well-Sweep Herb Farm, who offer both seed and
plants.

Asian Brassicas: Chinese Cabbage, Michihli, and Bok Choy

Brassica pekinensis; B. chinensis

THESE brassicas are so much alike in origin, history, and cultivation that they are often confused, and it may help simplify matters for us to deal with them together. They are all native to eastern Asia, and have been around for many thousands of years. Oddly enough, oriental brassicas did not make their appearance in Europe until the eighteenth century, and failed to become popular until late in the 1800s. The multitude of vegetables which are collectively referred to in this country as "Chinese cabbage" seems astounding. What with vernacular English names, Cantonese and other Chinese names, Japanese names, and variety names, it would seem that there are at least a dozen different vegetables.

Some of them are loose-leaf and some grow in tightly compact heads like cabbage. All of them can be great assets to American cooks. They are finer in taste than cabbage, although somewhat more delicate to handle. As an added advantage, they will not cause digestion problems as cabbage does for some people.

This seeming legion of green vegetables, when their many titles are distinguished one from the other, sort themselves into three main types. There are other oriental brassicas, of course, but these three are the most well-known in this country, and the most readily available from seed houses.

One of them, *Brassica pekinensis,* is most often called CHINESE CABBAGE because of its tight head. The Cantonese call this SIEW CHOY. In Japanese the name is HAKU-SAI. Some catalogs list it as WONG BOK (another Chinese name). Other seed companies call their version of it NAGAOKA (Japanese for "long hill"). On the West Coast you may also encounter the name NAPA. The origin of this appellation is most likely from the Japanese word *nappa,* a generic term for many leafy green vegetables. Despite the similarity in name, the only likeness Chinese cabbage really shares with our ordinary cabbage is its tightly curled head and blanched inner leaves. Most people feel the flavor is very different from that of common cabbage, but it does remind some of Savoy cabbage.

A similar vegetable is MICHIHLI or CHIHLI, which has long been cultivated throughout America and remains the most popular type today. It is a compact head like Chinese cabbage, but more slender and torpedo-shaped, somewhat similar in appearance to romaine lettuce. It is sometimes called GREEN CHINESE CABBAGE because the head is looser and greener than that of Chinese cabbage. Michihli, which is available from a number of seed companies, is supposed to be an improvement on the older Chihli, which is seldom seen in seed catalogs these days. Also called CHING SIEW CHOY in Cantonese, this type tastes something like Chinese cabbage, but stronger and more pungent. Michihli is sometimes called PEI-TSAI as well, but that name is also applied to Bok Choy. If you come upon it, look for a description to be sure which vegetable is meant.

The third brassica, which forms a loose-leaf head, is often called CELERY CABBAGE or CHINESE CELERY, but it is very different from either celery or cabbage in taste, in shape, and in cultivation. This one, *Brassica chinensis,* is actually a mustard. Like celery, it has thick, white stalks. Unlike celery, however, the stalks of Chinese celery are not fibrous. Rather, they are creamy white and very tender. Each stalk is actually the central stem of a leaf which is large, dark green, and spoon-shaped. This vegetable is most often called by its Cantonese name, BOK CHOY or PAK CHOI depending on how you spell it, but is also known as PE-TSAI or PEI TSAI (in Mandarin). By whatever name you choose to call it, this is a star among vegetables. It has a flavor and texture unto itself—tender, sweet, crisp, and milder tasting than either Michihli or Chinese cabbage.

Many varieties of these Asian brassicas are short-day-length growers. If planted late in spring the plants go quickly to seed. Plant breeders have now given us some varieties that don't possess this short-day character, and we can now get our brassicas to grow through spring, summer, and winter. You'll have to determine for yourself whether the seed you purchase is short-day-length sensitive or not, because often the seed packets and catalogs do not mention this problem. If you're in doubt, ask the seed company which kind they sell.

Like more familiar brassicas such as broccoli or ordinary cabbage, the faster the Asian types are grown to maturity the more readily they will develop to their characteristic size, and the better they will taste. This calls for extremely rich soil high in nitrogen, which can easily be obtained by heavy applications of well-rotted manure. These plants are heavy feeders, and will also need to be side-dressed with applications of fertilizer after four weeks of growth. Bi-weekly doses of liquid manure will pay off in larger, faster-developed plants.

So long as sufficient nutrients are available, the plants will tolerate just about any type of soil from tight clay to loose sand. They do need ample moisture; lack of it can retard growth and cause the plants to bolt. A sudden hot spell after a long period of cool weather can also cause these plants to bolt. Similarly, the shock of transplanting is enough to make the plants bolt prematurely.

Therefore, it is best to seed the plants directly in the garden and thin them later. Plant either as soon as frost is past in the spring (if the variety does not demand a short day length) or in mid to late summer for an autumn crop. Although the plants aren't usually regarded as frost-hardy, one Oregon gardener reported that his last crop came through 20-degree frosts with flying colors. Although all three of these plants can be cultivated in the same way, you will find that seed packets, catalogs, and gardening books all give planting directions that

• I plant Bok Choy in early spring here in central New Jersey. It is usable at any time after sprouting; but, of course, the bigger the better. So I let the plants grow as long as they are edible. One or two days of heat then makes them bitter, and causes them to go to seed.

As a fall and winter vegetable, Bok Choy is a must in my garden. I start seedlings in a shaded area in August, transplant them to the main bed in September, then transplant them once more to the cold frame in November. For space and extra protection I dig out the soil under the cold frame to a depth of 12 inches. In this way, Bok Choy not only lasts all winter; it even grows noticeably during slightly warm spells.

*—Kenneth Negus
Princeton, New Jersey*

• I have raised Michihli for about six years now. I have never had a failure. It is one of my "sure things" in the garden . . . It looked good so I bought a pack of seeds to try it.

I fertilized it with raw chicken compost and it grew so fast that I was afraid it would be spindly. It produced so many heads that I had plenty to share with neighbors who had never seen such a plant.

. . . One thing to remember about it is that it needs "growing room." It needs at least 6 to 8 inches in between the plants for it to be big and juicy. I found that if I neglected to thin it out properly it would be tall, spindly and tough.

Chinese cabbage is a beautiful plant in all stages of growth, especially just as it is heading. It reminds me of a bouquet of greenery wrapped up in a ruffled piece of green wrapping paper. If you couldn't eat it I guess I would grow it anyway as a flower.

—*R. L. Brewer*
Hampton, South Carolina

vary somewhat. As a general rule of thumb, plant the seed about ½ inch deep (less if your soil is heavy) and 3 to 4 inches apart, thinning back eventually to 10 inches apart, or 18 inches for Chinese cabbage. Thin the plants gradually over a period of time so that the cook in your household can take advantage of them. Thinnings can be eaten raw in salads or used just like the adult plants in stir-fry dishes and soups.

Rows are generally recommended to be spaced 2 to 2½ feet apart, but can be placed closer if your land is rich enough to support a more intense cultivation. About 20 feet of row mixed with the three vegetables will be enough to introduce a family of four to the delights of dining on these oriental brassicas.

Cabbageworms are fond of these vegetables, and you need to catch them right away if they take a fancy to your crop. Plantings of onion, garlic or tansy around the bed will help repel cabbageworms. You might also try sprinkling the plants with rye flour when the dew is still on them. The worms get all gummed up in the resulting dough, and are killed when they bake in the sun. Thuricide, or *bacillus thuringiensis*, works magnificently against these pests. Potato aphids may also sometimes attack, but are effectively controlled with a strong stream of water from the garden hose.

Ann Harman, a Maryland gardener, reported that flea beetles in her garden ignored everything else they usually eat and reduced her Bok Choy and Wong Bok to lacework. To spare yourself the same plight, if flea beetles visit your garden, you might try a garlic-water spray, or a spoonful of wood ashes sprinkled on each plant two or three times a week. Some gardeners report problems with root maggots as well, and these may be combatted with rotenone dusting early in the development of the plants.

Asian brassicas may be harvested when the heads get large and tightly curled. Chinese cabbage should weigh several pounds at maturity; Michihli and Bok Choy should weigh in at about 3 to 4 pounds. If your plants are smaller, it means you will need to fertilize more next year.

Chinese cabbage and Bok Choy both store extremely well in the refrigerator. Like cabbage, they will also keep for several weeks in a cool cellar. A somewhat more involved but equally effective method of storage is to dig the entire plants and, with soil still clinging to their roots, put them in a cold frame covered with straw and a mound of earth, or in an outdoor pit filled with damp sand. Remove the loose, outer leaves from the plants before storing them this way. They should survive several months.

Michihli, on the other hand, does not store too well. It can be refrigerated for a couple of weeks; however, if left lying on its side, the head will become "L" shaped. It tends to keep on growing a little even after being cut and refrigerated.

To make the best use of these vegetables, you will want to become familiar with Japanese and Chinese cookery, and perhaps learn a few new recipes. Chinese cabbage can be used raw in salads, cooked in soups, cooked alone as a side dish, or stir-fried with other vegetables and/or meat. The Koreans make a pickled dish called *Kimchee* from Chinese cabbage. Michihli, with its more pungent flavor, is used in many of the same ways. Bok Choy is not customarily eaten raw, but is delicious when added to soups or stir-fried with meat and/or vegetables. It is a standard ingredient of wonton soup.

One classic way to serve any of the three is to stir-fry bite-sized slices with tofu (soybean curd), meat, and seasonings. For a different taste, substitute them in your favorite cabbage recipes. Another simple idea is to mix the chopped leaves with sliced radishes and Jerusalem artichokes for a hearty salad. When you've become a graduate student of Japanese cooking, you can make delicious pickles with these vegetables.

Most seed companies supply at least one of these brassicas. W. Atlee Burpee Co.; Comstock, Ferre and Co.; William Dam Seeds; DeGiorgi Co.; J.L. Hudson, Seedsman; Johnny's Selected Seeds; Kitazawa Seed Co.; and Sunrise Enterprises all handle several varieties.

Here's a delicious way to prepare any of these oriental brassicas: Chop leaves (and stems as well of Bok Choy) on the diagonal into 2- to 3-inch bite-sized pieces. Chop an onion and saute it in oil, along with a slice of ginger root if you have it. Add the cabbage, stir-fry ever so briefly, and stir in soy sauce to taste and just a bit of honey. Serve the vegetables with steamed brown rice. If your gobo is ready for harvest at the same time, by all means saute some thin slices of it along with the onion; it adds a contrasting crunchiness to the succulent quality of the dish.

Broccoli Raab

Brassica campestris

THIS variously-spelled vegetable is well known to Americans of Italian stock, who look for it eagerly at the market during winter months. Despite its many partisans, however, broccoli raab is rarely mentioned in gardening books and such. What a shame, for this tender shoot of the wintered-over turnip is a welcome sight in early April, when it parts the soil as one of the first greens of a new gardening year. Also called turnip broccoli, which is almost a direct translation of the Italian phrase *broccolini di rapa,* broccoli raab appears in the gardens of many Italian-Americans, with at least some of them continuing strains brought from the old country up to 50 years ago.

This Italian affection for the juicy flowering stems of the turnip probably dates back at least to the days of the Roman Empire. Upper-class Romans of that time were tremendously fond of the delicate shoots that grew up in the spring from the stalks of early forms of the brassicas cabbage and broccoli. It's plausible to guess that the turnip was harvested in this way too, for we know that turnip roots were an important food of poorer Romans and of country people. Since Pliny reports that the Romans had a special relish for turnip greens that were yellow and wilted, it seems more than likely that the succulent stems were a luxury indeed, bringing good prices along with other relatively scarce brassica shoots.

65

• I consider broccoli raab a permanent fixture in my garden. Why not? The seeds are free, saved from a plant or two the year before. Since I plant it in open spots and in between tender plants, pick some of the tender leaves in the fall, and harvest it in time to get it out of the way for May planting, it seems to take up zero space. It does require some planning—if you were to make a late decision to plant carrots or other early seeds where you had planted broccoli raab in the fall you might feel a bit frustrated.

This is a very simple vegetable, and it can pretty well look after itself. If there is a little space I plant a row. A little more, a wide row. If there is a broad area, the seed is sown broadcast. All the plants don't make it through the winter, and there always seems to be room to step between the plants. The harvest is small on a per-plant basis. However, in the second week of April any harvest is something big.

—Charles McNamara
Paramus, New Jersey

As a very hardy biennial, the turnip produces a well-developed root and/or luxuriant leaves the first year. Then the leaves die back, and the root gradually gets less succulent and more woody as the plant begins to channel its energy into a flowering stem that will enable it to fulfill its destiny by going to seed. If the plant survives the winter, a pulpy and juicy stem with a tight seedpod at its tip emerges from the core of the root in early spring. This sprout grows to 12 inches or so before flowering, its tip resembling a miniature head of broccoli.

You can obtain broccoli raab from any kind of turnip, but it seems best to choose a variety grown for greens rather than for the root, since well-formed roots may heave up during the winter. If your cold season is pretty severe, also look for a very hardy turnip which will have a fair chance of wintering over successfully under lots of mulch. One variety that has performed magnificently in New Jersey is SEVEN TOP, which puts forth substantial first stems in early spring and obligingly responds to cutting by developing smaller side shoots that provide a second harvest.

Turnips can be grown anywhere (and anytime) you can provide a temperate climate, adequate moisture, and good drainage. For late summer planting, sow the seeds ½ inch deep—and as little as 2 inches apart—in space vacated by crops harvested earlier, and in between widely spaced plants that will be killed by frost. Broccoli raab likes soil with a pH around 6.5, and lots of nitrogen. Though turnips do well in relatively poor soils and will thrive on nutrients left over by the crop they follow, you'll get better results if you cover your seeds with sifted compost. Also water them thoroughly and make sure the germinating seeds don't dry out in the late summer heat. Your germination rate should be high, but don't bother to thin those seedlings since you're likely to lose some plants over the winter.

Winter survival aside, broccoli raab seems to be a pretty dependable crop, for planting late helps protect it from the maggots that infest brassicas in some areas. After you harvest your turnip greens and the first freeze hits, put down a heavy mulch if your growing area is too rough in winter for biennials on their own. Then keep your fingers crossed until the temperature begins to climb again and it's time to look for emerging spears under your mulch.

Harvest the flowering shoots while they're still tender and before blossoms appear. They'll taste best when they're 6 to 8 inches long, since the longer, older shoots will be thick and tough at the base and will have to be cut or broken off much like aging asparagus spears. You'll probably be able to harvest a second crop of smaller stalks, but the shoots that struggle out after that will be too small for good eating. When your plants start to flower, add them to the compost heap, leaving just one or two to provide seeds for the next planting a few months later.

Broccoli raab is delicious when cooked and served like asparagus, and you might want to try it with hollandaise sauce or lemon butter. It's also good creamed or au gratin. For a classic Italian treat, blanch the shoots by boiling them for a few minutes. Meanwhile saute a garlic clove or two in some olive oil, then remove the garlic. Add the wet stems to the hot oil and steam them covered for about 5 minutes. In the unlikely event that you're troubled by leftovers, enjoy the stems and their juice over vermicelli.

Some mail order sources for turnips yielding good broccoli raab are Comstock, Ferre and Co.; DeGiorgi Co.; Joseph Harris Co., Inc.; and R.H. Shumway Seedsman.

Domestic Burdock

Arctium lappa

IF you've ever set off on a stroll through a lush green pasture or cool, leafy forest only to come out the other side covered with prickly burrs tenaciously clinging to your clothes and hair, you've made the acquaintance of the wild burdock. As you're trying to pick the stubborn burrs off yourself or your dog, burdock probably seems like the last plant you'd want to have in your vegetable garden. But the Japanese have domesticated the plant, which they call gobo, or Takinogawa, and have grown it for years. Its roots are valued highly as both a food and a tonic to purify the blood and relieve arthritis. It is also said to help gout and skin diseases. Burdock roots do indeed lend a pleasantly aromatic flavor to vegetable dishes and soups, and so long as you don't allow them to go to seed, they won't develop their pesky burrs (which are actually containers for the seeds). The plant grows large, fan-shaped, light green leaves above the ground, and long, brown-skinned white roots below.

The large leaves are the only real drawback to growing gobo, unless you have lots of room. The plant itself does not deplete the soil around it of nutrients, nor does it spread out like an untrained tomato, but the huge leaves shade other plants, and can thus make it difficult to garden intensively in a small area.

Gobo thrives in just about any kind of soil between a swamp and an arid mountaintop (which is why it's such a pest

69

• Our gobo is grown for my husband's parents, who cannot grow it due to rocky soil. Gobo needs good drainage and ample moisture and nutrients. Our soil is slightly sandy and works fine. The first year we planted it, we dug a trench about 18 inches deep and a shovel wide. We felt this would be adequate. Our first crop was so-so. Then we discovered the secret (for us)! After digging out the gobo (it has to be dug or else you can't harvest the entire root) we left the trench due to rains. As we accumulated leaves, kitchen trimmings, etc., we would throw them into our trench and put down a layer of dirt on top, making a gigantic compost trench. Come springtime the trench had been filled in and was ready for planting. The crops thereafter have been plentiful. Our trench has been enlarged to where it is 3 to 4 feet wide and 2½ to 3 feet deep. Our trench is 20 feet long and our planting row is about 9 inches wide. We will continue to grow gobo as it is very expensive (if it can be found) in the store.

—*Connie Nakagawa*
Fair Oaks, California

where it grows naturally). But if grown in loose soil, the roots will grow straighter. Plant the seeds in spring when the ground has warmed. Dig in the seed ¼ inch deep, and from 4 to 8 inches apart. Burdock needs little care until harvest time. If your soil is exceptionally poor, some compost or fertilizer might be in order, but otherwise your burdock will probably thrive on its own.

Burdock doesn't need much water; if you water it too much, you fail to stimulate root growth because the root will not move downward in search of water. Too much water in the early stages of growth also makes the leaves more prone to insect attack. A good grass mulch around the burdock will help to conserve water, and make all but occasional watering unnecessary. Grasshoppers seem to be the only insects that are fond of gobo leaves. Nothing else seems to bother them, not even weeds.

Burdock grows well in greenhouse conditions, as well as outdoors. Seeds may be planted directly in tubs or in the bench, or you can start them in peat pots to save space. Transplant the seedlings when they are 4 to 5 inches tall, to stand about a foot and a half apart. If your benches aren't deep enough to allow room for the roots, you can invert some bottomless boxes on the bench, fill them with soil, and plant in them.

The problem with gobo is not growing it, but digging it up. Anyone in New England with a stony field full of burdock will say Amen to that. The best way around the problem is to plant the seed in soil that needs a deep digging anyway to put it in shape, or soil you won't mind digging because it's soft and loose. The roots go down 2 to 3 feet; in fact, given the opportunity, burdock will grow to the size of a baseball bat, so to keep gobo from becoming a pest, you really need to get most of it out. If the tip of the main root, or even a large feeder root, snaps off and is left in the soil, it will sprout. Burdock is usually harvested from three to five months after planting. It is best if harvested when still young and tender. If left to fully mature, the flavor is still delicious, but the skin of the root becomes tough and must be peeled. When the leaves have gotten quite large in the fall, the roots are probably big enough to use. In mild regions, gobo can be left in the ground after the leaves die back, and dug as needed. Gobo should be dug during the first year because the plant is a biennial, setting seedheads and getting quite tough the second year.

Here's a good method of digging gobo: Alongside the row of burdock, dig a trench 2½ feet deep. This will open up one side of the roots to your hands. Instead of pulling straight up on the roots, pull at a 75-degree angle toward the trench. The roots should come easily, and will be less likely to break.

To store gobo, cut off the tops, wash the roots, and dry them. They store fairly well in the refrigerator if wrapped in damp cloths. If the cloths are refreshed and aired, the roots will probably keep for a couple of weeks. To store gobo for a longer period of time, wrap the roots in newspaper, then put them in a plastic bag and fasten it tightly with rubber bands. The roots should last several months if kept in a cool place. You can also bury the roots in a pit of clean sand covered with boards or plastic.

When you're ready to prepare the roots for eating, scrape the outer layer of the roots down to the white fiber beneath. The skin scrapes off easily. Slice the roots in thick julienne strips or cut them on the diagonal. The pieces may either be soaked in cool water or go directly to the cooking pot. Either way, it's a good idea to change the soaking or cooking water two or three times to improve the vegetable's color and flavor. In Japan, the roots can be purchased already sliced and peeled, from tubs of water where they are left to soak until the customer wants them.

Cook gobo until tender—its refreshing, sweetish, unusually aromatic flavor develops with thorough cooking. When cooked, gobo is crunchier than a cooked carrot, and slightly stringy, though not unpleasantly so. Use it as an ingredient in oriental stir-fry dishes, combine it with carrots for a side dish, or add it to soups to impart a really interesting and unusual flavor. It can be chopped very fine and used as a condiment with bland rice or fish dishes, or deep-fried. Some people say that when the roots are wrapped in wet heavy paper and roasted over hot coals, they taste surprisingly like meat.

Seed for edible burdock is available from Kitazawa Seed Company and Nichols Garden Nursery.

• A problem for home gardeners is that burdock volunteers can get out of control and sprout all over the garden. Many Japanese farmers build a special box raised several feet above the ground and filled with soil, for growing burdock. When the root is ready the box is dismantled, saving the work of digging deep into the soil to harvest, and also serving to keep the burdock under control.

Burdock is a popular vegetable in Japan. It is boiled and eaten in soup, cooked as tempura, and fried with vegetables and seaweed. It can be substituted for carrot in almost any recipe, but in Japan it is never eaten raw.

—Larry Korn
San Francisco, California

Salad Burnet

Poterium Sanguisorba or *Sanguisorba minor*

WHEN cooks and gardeners begin praising salad burnet, they are usually referring to the perennial herb which taxonomists label *Poterium Sanguisorba* or *Sanguisorba minor*. It is a hardy, sprawling plant with white, pink or purplish flower spikes, bright red seedheads, and scallop-edged leaves that are arranged in pairs along the stem. The leaves have a distinctive cucumber-like flavor and are harvested when young for use in assorted foods, beverages, and medicinal preparations. Great burnet (*Sanguisorba officinalis*) can be used in all of these ways, but it remains primarily a wild, meadow herb. It differs from salad burnet in having more deeply colored flowers of a slightly different formation.

Both burnets are native to Europe and western Asia and have been naturalized in North America. Members of the rose family, they are said to have been known and perhaps even cultivated since ancient times. Early Anglo-Saxons used burnet as a key ingredient of their green-salve, as well as in certain other medicines and ointments. During the fifteenth through seventeenth centuries, burnet was used to heal wounds, cure gout, and protect people against the plague. Soon it became an important foodstuff as well. The early colonists brought the seeds to this country, where burnet has since flourished in both field and garden. The herbalists Gerard, Evelyn, and Parkinson

73

all describe it as a most beneficial and useful medicinal herb which is especially fine when added to a glass of wine. Thomas Jefferson lists it as a major "sallet green" of his kitchen garden.

Burnet can be grown as an annual or perennial almost anywhere in the country. It is very hardy, and will remain green year-round in most climatic zones, producing greens continuously except when snowfall is very heavy or the weather particularly bitter. Burnet grows readily in almost any soil as long as full sunlight is provided. Indeed, it seems to prefer dry, poor soil and is at its healthiest and best-tasting when its bed has not been fertilized or kept constantly moist. Full sunlight is important, however, as is a fairly neutral soil. Very acid ground should be treated with wood ashes before seeding.

Plant burnet seeds outdoors about the time of the last expected spring freeze. You can either broadcast seeds and cover them ⅛ inch deep, or sow them thinly in rows 1 foot wide. Germination takes place in about ten days and, several weeks later, the seedlings can be thinned to stand 8 to 10 inches apart. For a particularly lush, dense patch of this decorative herb, plant seeds just a few inches apart and do not thin.

If burnet is grown as an annual, allow the plants to self-seed before cutting down the top growth at the end of the season. Otherwise, cut back most of the plants just after their June flowering, leaving only a few to produce and scatter seeds. New growth will follow; the mature plants will continue to provide a supply of garden greens all year long, and young seedlings will appear the following spring. If well protected with a winter mulch, fresh leaves can be had even during the coldest months of the year. The following spring, new plants will arise from the self-sown seeds, thereby replacing those older plants which die during the winter. Although some gardeners recommend dividing and replanting burnet roots each spring, this is both difficult and unnecessary.

As they grow, the plants send out very low, spreading branches which stay close to the ground until flowering time. When the flower heads appear, the foliage rises to a height of about 12 inches. Leaves can be picked at any time after the plants have become established and have developed several branches. Freshly harvested burnet is far superior to any dried or frozen leaves. Pick them just before you're going to eat them. In salads, they add a refreshing, delicate cucumber taste to other greens. Cooked in cream soups such as asparagus or mushroom, they are especially delicious. A fresh sprig of burnet makes a very pretty and charming garnish for a cup of hot or cold soup. It can be used instead of dill to garnish casseroles and fish dishes. Or, try chopping the leaves and mixing them with softened butter for a gourmet sandwich spread. But, while these methods of serving burnet are all very worthwhile and even superb, the herb is only, as Evelyn claimed, in "its most genuine element" when served in a cup of claret. Once a major ingredient of ale, burnet can be steeped in wine or simply served fresh in a cool summer drink.

Seeds for salad burnet can be purchased from Casa Yerba; Comstock, Ferre and Co.; William Dam Seeds; J.A. Demonchaux Co.; J.L. Hudson, Seedsman; Meadowbrook Herb Garden; Geo. W. Park Seed Co.; and Redwood City Seed Co. Plants are available from Hemlock Hill Herb Farm, and either seeds or plants from Well-Sweep Herb Farm.

• The leaves have a delicate fragrance and taste like cucumbers, which when added to sandwiches or salads brings compliments and questions from both company and family members. You could be smug and let them think you are really growing cucumbers out there in the cold of winter, but I simply can't resist telling them about salad burnet and giving them a few plants to take home for their own enjoyment.

—*Nancy Alice Fisher*
Monmouth, Oregon

Nopal or Prickly Pear Cacti

Opuntia and *Nopalea, spp.*

OPUNTIA and the closely related *Nopalea* cacti are known to most American gardeners as hardy, rock garden ornamentals and to western farmers as pesky weeds. However, in many tropical and desert regions of the world, they are highly valued for their edible fruits and "leaves." Most edible forms of these genera are characterized by their flat, padlike stems which, when peeled and cooked, are eaten like okra. These are called *nopalitos* or, improperly, "leaves." The true leaves are very small and drop off early in the plant's development. Flowers are generally white, yellow, or orange, and are held to the pads by pear-shaped, edible fruits. The fruits range in color from whitish to yellow and, more commonly, red or purple. Their shape and prickly skin have given them the name "prickly pears." Known also as *tunas*, they are usually eaten fresh like melon, but are sometimes cooked and fermented to make a beverage. Horticultural researchers at Texas A & M University have lauded prickly pears, with their high vitamin and sugar content, as a potentially important source of food that's been too long overlooked.

The two genera, *Opuntia* and *Nopalea*, differ only slightly and some taxonomists fail to distinguish between them. Indeed, the common name of many *Opuntia* is nopal or nopalillo. Several of the *Nopalea* are known as prickly pears. Technically,

• Vendors in the streets of old Mexico used to have prickly pears cooled on blocks of ice in their carts. When one was purchasing a *tuna*, the vendor would expertly impale the prickly pear on a clean stick and, rotating it rapidly in one hand, he would clean off the outer skin and thorns with a few flashes of a very sharp knife. *Tunas* taste very much like watermelon. They are very refreshing and sweet. On a hot day they are more welcome than those celebrated drinks of carbonated water.

The drawback is that *tunas* are about half seed and half melon-like flesh. They have only a little better proportion of flesh to seed than a pomegranate. The connois-

the *Nopalea* genus includes only three edible, cultivated species: *N. dejecta, N. chamacuera* and *N. cochenillifera*. Many consider the latter the true "nopal" and one of the oldest of cultivated cacti. Although the fruits are eaten locally, this type is grown primarily as food for a certain type of mealybug from which a red dye is extracted. *N. dejecta* and *N. chamacuera* are grown strictly for their deep red prickly pears, which, along with the thick stems, form a sizeable portion of the Mexican diet.

Edible *Opuntia* include *O. engelmanii* or *O. phaecantha,* which is a low-growing prickly pear with oval pads 4 to 16 inches long, 9 inches wide and about ¾ inch thick. These pads are fried and eaten by Mexicans and Southwestern native Americans. *O. leucotricha* is a much taller, tree-like *Opuntia* with 4- to 10-inch pads, yellow flowers, and white to red aromatic fruits. Both the stems and the fruits are eaten and enjoyed throughout Mexico and in California and Arizona. A third *Opuntia, O. basilaris* or BEAVER-TAILED CACTUS, is grown for its flowers which are steamed in pits dug in the ground, and prepared as one of the first spring foods.

By far the most popular edible prickly pear cactus is the INDIAN FIG. Classified variously as *O. megacantha, O. occidentalis,* and *O. ficus—indica,* it produces yellow flowers and deep red or whitish pears which are said to be the best of all the *Opuntia.* Plants vary in size; some grow prostrate, while others reach heights of 6 or 10 feet. The pads and fruits have relatively few spines. Several popular varieties of prickly pear, such as BURBANK'S SPINELESS CACTUS, are completely spineless and require little or no cleaning to prepare for table. They bloom throughout the summer, providing a continuous supply of fruits.

All of these cacti share the same cultural requirements. Although native to the Southwest, and to Central and South America, many will thrive as far north as Massachusetts and eastward to the Atlantic seaboard, as long as winter protection is provided. In these regions, the cacti should be grown in containers that can be moved indoors in winter. Or, plant them in a sheltered area beside a building; the warm soil beside a greenhouse is ideal.

In regions to which the cactus is native, seeds will sprout readily if simply scattered in a sandy garden bed. Elsewhere, the best procedure is to sow them indoors in flats. Broadcast seeds thinly in a bed of moist vermiculite, peat, and sand, or very light sterile soil. Sprinkle a little of the medium over them and water very lightly. If the flat is covered with plastic or a pane of glass, the medium will not dry out and will require no more watering. Dry the glass as large water droplets appear. On sunny days, cover the glass with a single sheet of newspaper to prevent roasting of the seed.

Growth from seed is rather slow, and several years are required before the cacti are large enough to flower and bear fruit. Vegetative propagation is simpler and faster. Simply cut a pad or a branch with many joints from a growing cactus and plant it in well-drained garden soil. A mixture of half soil, half sand, and a bit of crushed pottery, leaf mold, and bone meal will produce excellent results. Plant the pad several inches deep in the soil; only one-third of it should be above ground. Again, keep the rooting medium moist but not wet. The moisture stored within the plant's tissues will enable it to sprout roots without extra water. After two months, the cutting will stand firmly in the soil and a stray root or two may be visible in the drainage holes. If the pad is planted in early spring, roots and new pads should appear within several months.

Once the roots have developed, the cactus will grow in practically any soil. Rocky and even clayey ground can support the plants as long as fairly good drainage is provided.

These cacti are extremely easy to grow and will not need pampering. Even when summers are totally without rain, they do not need water. Too much moisture causes them to mildew and eventually to rot.

Small pads or nopalitos can be picked and eaten at any time. They are slightly chewy, very fresh to the taste, and mucilagenous, similar to okra. To prepare them, simply scrape the skins with a blunt knife until all the spines have been removed. Peel and dice the cleaned nopalitos and boil them, with garlic and onions, until tender. Prepared in this way, they are especially delicious with shellfish, pork, or eggs. Try them scrambled with eggs and chile sauce for a new taste at breakfast.

seur of *tunas* has two alternatives: swallow the seeds or hold them in the mouth while chewing off and eating the sweet flesh, saving them to be spit out later. Because the seeds are about the size of double-ought gunshot, swallowing them stimulates a certain amount of introspection or at least some reflection on the wonders of the gastrointestinal tract. Spitting out the seeds of the *tuna* on the street is not likely to appeal to the fastidious. I chose a "chicken-Yankee-compromise." I spit them into my hand and threw them into the gutter when I thought I was least conspicuous.

—*John Meeker
Gilroy, California*

The real glory of the cactus is the pear or *tuna*. These appear in midsummer and continue until the first frost. Sold by street vendors in Mexico and Israel and available in grocery stores throughout California and the Southwest, they are very refreshing and as sweet as many melons. Pick the fruits when they have turned deep red or yellow and begin to shrivel up a bit. Put on a pair of heavy gloves and rub the fruits to remove the prickles, or singe them over a flame for a moment. You can serve them chilled and sprinkled with lemon juice, or cook them to make a syrup (*mielde tuna*). This syrup can then be boiled to make a sweet paste (*mel coacha*), or fermented into a drink. Dried prickly pears are sometimes made into a flour from which small, sweet cakes are made (*queso de tuna*). The famous "cactus candy" is made from the cooked pears of the *Opuntia* that have been mentioned here.

Seeds of the Indian Fig variety are available from Exotica Seed Company.

Marianka de Nijs of Corona, California recommends this method of preparing nopal pads:

Combine 1 cup freshly cooked nopales with 1 cup chopped fresh tomatoes, 1 teaspoon chopped coriander, one fresh chile pepper, peeled and minced, and about ¼ teaspoon salt. Toss and chill or toss and heat through.

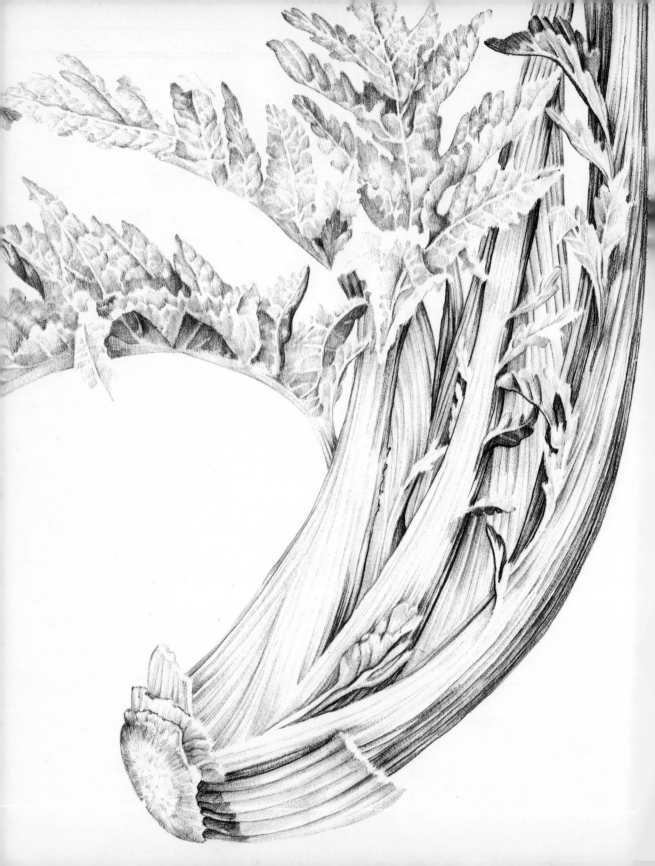

Cardoon

Cynara cardunculus

THOUGH it sounds more like a sweater made in Scotland, cardoon is the ancestor of the globe artichoke, taking its name from the French word for thistle (*chardon*). Originating in southern Europe and north Africa, this showy tender perennial may well have been the first plant under human cultivation, for seeds have been found that were gathered more than 30,000 years ago! The Romans relished cardoon's huge leaf stalks so much that they imported them from Libya and Spain in massive amounts. And to this day cardoon is considered choice eating around the Mediterranean—particularly in Italy, where it's called *cardoni,* and in France, where several varieties are painstakingly cultivated. So far, though, cardoon is little known and even less grown in America, probably because its somewhat bitter flavor is an acquired taste and because the culture and preparation of this striking vegetable are a bit tricky and time-consuming. Nevertheless, gardeners who grow cardoon successfully and serve it in appetizing ways rave over its subtle flavor and unusual texture, and wait eagerly for it to come into season.

Cardoon is highly ornamental and even somewhat spectacular, reaching up to 6 feet in height and almost as much in width. The plant looks a lot like the globe artichoke—in fact some botanists classify the two as varieties of the same species.

Cardoon, however, usually has spinier leaves, and its flower buds don't form edible heads. Instead, the plant's energy goes into huge crisp chards that may be eaten green when young or blanched much like celery as they race toward maturity. Most kinds of cardoon run to seed sooner or later, sending up lovely purplish blue thistle blooms from amidst large grey green leaves that are deeply cleft and felted. When these spikey seedheads have dried in the autumn, they are much sought after for dried flower arrangements.

Cardoon needs a long growing season, but it can be cultivated as an annual in more northern parts of the United States if you start it early indoors and set it out in late May. (Indeed, some people say you get better-tasting plants this way.) The season can be stretched even further by lifting the plants roots and all and blanching them in a root cellar or similar place for the last month of their growth. In southern California and other places where the ground never freezes hard, cardoon will send out new growth in the spring if its roots and crowns are left undisturbed. Where this happens, new plants can be propagated by dividing the roots or working with suckers.

Several varieties of cardoon have been developed in Europe, where it's much fancied. SPANISH grows quite large and has the advantage of spineless leaves, but tends to go to seed early. The French prefer TOURS, a variety that is reasonably hardy, produces thick, solid, tender chards, and is less likely than Spanish to go to seed quickly. Unfortunately, although Tours has tasty ribs it also has very spiny leaves which make it hard on a gardener's hands. A third choice is MAR-SEILLES, a red-stemmed cardoon with long leaves and solid stalks of good quality. The favorite of many, however, is PARIS, a large plant with few spines and stems considered the tenderest and best-tasting of all when cooked.

In California and the mildest parts of the South, you can sow cardoon seeds directly in early April in a foot-deep trench to which you've added several inches of compost. Plant two or three seeds at 1-foot intervals, covering them with an inch of soil. Thin the seedlings to 3 feet apart when they are big enough to handle. In cooler climates, sow in mid-March in a gentle hotbed or in pots in the greenhouse. Transfer the plants to trenches outdoors at the end of May.

Cardoon likes rich, not-too-acid soil with very good drainage and generous amounts of compost or well-rotted manure. Though slow to get started, the plant will grow rapidly from midsummer. But keep the cardoon well watered or the harvested stems will be a hollow, woody, bitter-tasting disappointment. To protect yourself from such a catastrophe, you can dig a basin around each plant and soak it thoroughly once a week. Once in awhile use liquid manure to build big, juicy stalks. As another safeguard, put down a thick mulch.

You can begin cutting the stalks as needed when they're 12 to 24 inches high and still tender and solid. Cut or break off the young stalks close to the ground and discard the upper portion where it becomes too narrow to be useful. Then carefully trim off all leaf stems before cooking, for even the smallest leaflet will give a bitter taste if it's not removed. Cardoon of this age keeps well if you don't allow it to dry out, and may also be boiled, then canned.

If your cardoon grows faster than your appetite for it, you can make more mature chards white and edible by blanching your plants when they're about 3 feet tall. Just tie up the drooping leaves with cord, then wrap straw, burlap, or heavy brown paper around each plant and hill some earth up around its base. The stalks will be ready to eat a month later. They can be kept through the winter for use as needed by lifting them roots and all and storing them in sand in a cellar or shed where the temperature doesn't drop below the upper forties. If you get a late start with your cardoon or need to extend your growing season, you can even dig up some medium-sized plants and stand them in a box of earth in the cellar to blanch for a month. If the light is weak enough, they will self-blanch and you won't have to wrap them.

You'll find cardoon to be a versatile addition to your vegetable repertoire. For a delightful accompaniment to meats, try it prepared like French fries. Another method is to cook cardoon until tender in chicken broth, then drain and marinate in Italian salad dressing. This makes a surefire conversation piece on an antipasto platter. Or you can dress cooked cardoon with oil and lemon juice *a la grecque.* Or add it to a tossed salad. For a change, try cooked cardoon hot, with a cream or Mornay sauce or with herb butter. (It's a good idea to add a little lemon juice in the cooking water to keep the slices from darkening.) Because of its artichoke-like texture, this cooked vegetable also tastes fine in soups and stews and Mediterranean-style main dishes. It's also good pickled or deep-fried as tempura.

You can buy seed of the challenging cardoon from Casa Yerba; Comstock, Ferre and Co.; William Dam Seeds; DeGiorgi Co.; Gurney Seed and Nursery Co.; Charles C. Hart Seed Co.; J.L. Hudson, Seedsman; Le Jardin du Gourmet; Meadowbrook Herb Garden; Nichols Garden Nursery; Redwood City Seed Co.; R.H. Shumway Seedsman; Thompson and Morgan, Inc. (where it's listed as Cynara in the flower part of the catalog); and Well-Sweep Herb Farm.

• Cardoon has always been a family favorite and my mom has had good success in growing it. She makes it a point to can a lot of it each year. It is delicious dipped in flour and egg and fried or added to Italian dishes like chicken cacciatore and stews. . . .

Around September she ties the plants together with string to blanch for about three weeks. Then, as the weather begins to cool, she bends the plants gently to the ground and covers them with dirt, straw, or both, then leaves them another two weeks or so. She says this process makes the plants more tender. However, leave some air space when covering the plants or they will rot.

—*Mrs. L. Nottoli and*
Susan Wade
Hammond, Indiana

For some truly gourmet eating, cover stripped and cleaned 2- to 4-inch chunks of cardoon with salted water. Boil them until tender—this may take as long as 40 to 50 minutes if your cardoon is unblanched. Meanwhile, make a deep-frying batter by combining ¼ cup milk, 2 tablespoons whole wheat or unbleached flour, one egg, and a little salt. Add a handful of grated Romano cheese and a little crushed garlic. Dip the drained cardoons in this batter, then fry in a generous amount of olive oil until brown and crusty.

Purple Cauliflower/
Purple Broccoli

Brassica oleracea, var. *botrytis*

THE purple vegetables have always interested gardeners because they are reputed to be resistant to insect pests, and because they have that mysterious ability to change their color from purple to bright green during the cooking process. Purple cauliflower and broccoli have both these desirable characteristics.

These two vegetables in purple are perhaps the same vegetable. Even though they are advertised by different names—SICILIAN PURPLE CAULIFLOWER, KING ROBERT PURPLE BROCCOLI, EARLY PURPLE SPROUTING BROCCOLI, PURPLE HEAD CAULIFLOWER, and EARLY PURPLE HEAD, for example—you may wonder why the seed companies call one broccoli and the other cauliflower when, if grown under similar conditions, they all look the same. It seems to be just a matter of preference on the part of the company.

Whether the version you choose to grow is called broccoli or cauliflower is really unimportant. Both purple vegetables are hardier and easier to grow than either green headed broccoli or white cauliflower. The heads don't need to be blanched like those of white cauliflower. The vegetables taste good, and their resistance to pests means little extra work beyond fertilizing. The central head formed is a bit looser and smaller than that of white cauliflower, but is usually well developed. The lateral

89

shoots grow rapidly after the central head is cut, and if these shoots are cut as they mature, the plant will continue producing well into autumn, as long as frosts are not too heavy.

Many gardeners have found purple cauliflower and broccoli to withstand warm weather better than their white and green counterparts. Both mature in about 90 days, so you may be able to make two plantings, in April and again in July, if your summer temperatures aren't too severe. However heat-tolerant these vegetables are, though, remember that they like cool weather best. To be on the safe side, try to plan your planting dates so that your crop matures within three months, avoiding the heat of summer. Planting your first crop in March will let you harvest before summer gets into full swing. Delay your second planting until late July or early August, and you'll be able to harvest again in the cool of autumn, providing your autumns aren't too cool. Purple cauliflower and broccoli can take some light frost.

For your initial crop, start seed early in the season in small pots and transplant the seedlings out in the garden when the weather welcomes them. If properly hardened off, the young plants can resist considerable frost. Then, when you set out these transplants, sow some seed directly in the ground. This method should give you two harvests before the weather gets really hot.

Plant seeds ½ inch deep in loose soil, ¼ inch deep in tight soil. Space the young plants 16 to 18 inches apart. Make sure that the soil is extra rich in manure or other fertilizer, for these cauliflower/broccoli plants are heavy feeders. It's also good practice to provide manure tea or side-dressings of fertilizer through their early growing period at two- to three-week intervals. Fish emulsion is excellent to use as a base for the applied fertilizer. Like the other cole crops, without plenty of fertilizer, these plants will not develop the lovely large heads that the commercial growers get. Also, avoid the temptation to put the plants too close together—a mistake many gardeners have made in attempting too much in a small space—the plants will overly compete for the available fertilizer and produce inferior-sized heads.

• After harvesting the big heads of the delicately flavored purple cauliflower, don't throw away all those stems. The core of the stem is a tender taste treat. To use, cut the stem into manageable pieces and peel the fibrous cover off, leaving the white center. Raw, it tastes much like kohlrabi and can be used in salads or on the relish tray. Cooked, it is more like a mild white turnip, but with a nicer texture.

—Catharine M. Teall
Dundee, Michigan

Purple cauliflower and broccoli are harvested in the same way you harvest green broccoli. The central head is cut as soon as it's fully developed, while the buds are still tight. The smaller side shoots are cut as they become ready.

Both these vegetables freeze well for storage, and they adapt themselves easily to all your favorite broccoli and cauliflower recipes. Their excellent flavor is richer than that of white cauliflower, but milder than that of green broccoli. You may find reluctant children more willing to eat these vegetables than either of their namesakes. Try them with cheese sauce, lemon butter, or mustard sauce. Top them with almonds sauteed in butter, or sour cream and chopped hard-boiled eggs. Purple cauliflower and broccoli are very high in vitamin C, so by all means try them fresh from the garden in a salad.

Seed is available under one name or another from many companies, including Burgess Seed and Plant Co.; W. Atlee Burpee Co.; Comstock, Ferre & Co.; William Dam Seeds; Farmer Seed and Nursery Co.; Henry Field Seed and Nursery Co.; Joseph Harris Co., Inc.; Charles C. Hart Seed Co.; Nichols Garden Nursery; Geo. W. Park Co.; R.H. Shumway Seedsman; Stokes Seeds, Inc.; and Thompson and Morgan, Inc.

Celeriac

Apium graveolens var. *rapaceum*

THIS biennial belongs to the same species as celery, which originated in the Mediterranean area and seems to have started its long career as a medicinal herb and flavoring. Thought to be the *selinon* mentioned in Homer's *Odyssey*, primitive wild forms of celery are found in Europe, Asia Minor, and in the area between the Black and Caspian Seas, reaching over toward the Himalayas. Though this vegetable was mentioned in Chinese writings of the fifth century, it took another 1200 years before the gardeners of Italy, France, and England started to tone down its strong flavor through selective breeding. One of the improved varieties developed around this time was celeriac, which became a common European vegetable around 1700. Though still little grown in England and America, to this day celeriac ranks with sauerkraut and potatoes as a standard winter vegetable in Germany and eastern Europe.

Celeriac is sometimes called knob celery, turnip-rooted celery, or celery root, for it is grown not for its stalks and leaves, but for its globe-shaped root crown, which looks rather like a turnip. Having rough brown skin and whitish flesh, this solid and tender stem base consists mainly of the embryo of the plant but is also part taproot and stem. In flavor it is a sweeter and starchier version of celery, with a hint of parsley added. Not a root crop in the botanical sense, celeriac grows about 1½ feet high and about 1 foot across, with the tuber-like root crown reaching 4 inches or more in width. The leaf stalks of celeriac are shorter than those of celery and not as swollen, and its deep green foliage is more sparse and bitter.

Taking about 120 days to mature from the time the plants are set out, this variety of celery is a hardy cool-weather crop that can resist pretty stiff frosts. It thrives anywhere and anytime you can provide a mean temperature of 60° to 65°F. during the growing season, but won't do well if the average is over 70° to 75° or below 45°.

European seedhouses offer such varieties as APPLE and EARLY PARIS, but the celeriac most popular to date in America is LARGE SMOOTH PRAGUE. ALABASTER is a newer kind that some gardeners have found produces more uniform, generally larger bulbs than the Prague.

For best results, you should give celeriac the same culture as you would celery. Happily, though, it's less demanding about moisture and fertilizer—and it doesn't need blanching! If winters are very mild where you live, you can sow the seed directly outside in late summer, covering it with ⅛ inch of fine soil and with a board to keep the soil moist until germination. In areas with temperate summers, most gardeners sow very early in flats, cans, or pots. If this is your situation, get started in February or early March in a hotbed, late March on a windowsill, and early April in a cold greenhouse. And be sure to provide as much light as possible. Like other plants in the parsley family, celeriac has a low and slow germination rate, so you might use a generous amount of seed and hasten things along by placing the seeds on wet paper towels in a shallow glass- or plastic-covered pan and keeping them at a temperature of 70°F. for a few days. When they begin to sprout, mix them with cornmeal for easier handling and sow immediately ⅛ inch deep in organically rich earth, covering with fine soil.

Next, set your flat or can in water, but don't allow it to come in over the top. When the soil is thoroughly saturated, remove your planting container and place it in good light. If you haven't presoaked the seeds, you can hasten germination at this time by keeping the planted seeds at 60° to 65°F. at night and 10° warmer during the day. To help do this, cover the flat with a wet burlap sack. You can water on top of this or soak it and replace it when the flat gets dry. Or you can use glass or newspaper as a cover to help keep the soil moist and warm. But be careful to remove any cover just as soon as the young plants emerge or they will become spindly.

When your plants are 2 inches tall or have their true leaves, transplant them to another flat and space them 2 to 3 inches apart both ways. At 6 inches tall, they should be blocked to minimize planting-out shock. With a sharp knife cut between the rows one way, and in a week cut the remaining rows in the opposite direction. Harden the seedlings off by placing them outdoors on warm days before you set them out about eight to ten weeks after sowing. Plant 6 inches apart in rows 12 inches apart with the bulging base at ground level. (NOTE: choose your moment for planting-out carefully, for night temperatures below 45°F. can cause celeriac to put forth seed stalks instead of a rounded stem base. It's a good idea to wait until after the apple blossoms have fallen.)

Celeriac dotes on loose loamy or sandy soil, rich in compost and in potassium, and does nicely companion-planted with leeks or Scarlet Runner beans, which also like potassium. A previous sowing of a winter vetch also can help your celeriac: why not choose the fava bean, which will give you a valuable crop as it enriches the soil?

To help your celeriac root crowns burgeon beautifully, give each seedling a cup of liquid manure on transplanting and once a week thereafter. Hoe between plants lightly once in a while to keep down pest-harboring weeds—better yet, mulch, then water when needed, for if this vegetable dries out it will toughen. When the celery knobs have formed, trim off some of the lateral roots near the surface of each plant to keep them from making the fleshy stems of your celeriac oddly shaped and coarse. Then hill soil up around the swellings for some blanching during the last weeks of growth.

You can dig or pull up your celeriac when the globes are 2 to 2½ inches wide, which will probably be from late September to May. The knobs will keep for up to six months in a root cellar if you trim off the tops and the branching roots and store the crowns in moist sand as you would root crops. Celeriac stores best at below 40°F. but above freezing. As this suggests, if your winters are mild and your garden doesn't freeze hard, you can leave your mature celeriac in the ground over the winter, where it will stay firm and acquire a sweet and nutty flavor. To store it in this way, heap the soil in ridges over the plants and mulch them heavily with leaves and straw.

Though not a standout nutritionally, celeriac is a low-calorie vegetable that's ideal for weight-watchers. Many people find that once they've tried it they prefer its rich, mellow flavor to that of regular celery. And it makes a good home-garden substitute for celery since it often can be raised successfully where celery would fail and needs less tender loving care. Once this "lazy man's celery" has been washed and peeled, it can be used just like celery or turnips. While the stems and leaves are too bitter to eat raw, they make an excellent flavoring for soups and stews, and can keep you from having to buy expensive supermarket celery all summer and fall. The knobby, rough-skinned root is hard to peel; it's generally easier to slice the bulbs first and then peel the slices. Quartered, sliced, or cut julienne-style, celeriac root can be braised in broth, or boiled in salted water until it's tender and served with cheese, hollandaise, or cream sauce. Try it fried in butter until it's almost brown, or puree it and blend with mashed potatoes, as Europeans do. Celeriac also tastes fine in soups and stews or sliced thinly or grated into the salad bowl. Some folks marinate it in vinaigrette dressing, while others like it in fritters or cooked and pickled like red beets. You can also use the fresh leaves sparingly in salads or dry them to add celery flavoring to winter dishes.

Seed sources for celeriac include W. Atlee Burpee Co.; Casa Yerba; Comstock, Ferre and Co.; William Dam Seeds, DeGiorgi Co.; J.A. Demonchaux Co.; Henry Field Seed and Nursery Co.; Gurney Seed and Nursery Co.; Joseph Harris Co., Inc.; Charles C. Hart Seed Co.; J.L. Hudson, Seedsman; Le Jardin du Gourmet; Johnny's Selected Seeds; Meadowbrook Herb Garden; Nichols Garden Nursery; L.L. Olds Seed Co.;

Redwood City Seed Co.; R.H. Shumway Seedsman; Stokes Seeds, Inc. and Thompson and Morgan, Inc.

97
Celeriac

BRAISED CELERIAC

Here's a simple but delicious way to serve celeriac: wash and peel 6 to 8 celeriac knobs, and cut them into thin lengthwise slices. Saute the slices in a tablespoon of oil or butter in a heavy pot. Add 1 cup of vegetable, chicken, or beef stock, and bring to a boil. Cover and simmer until the celeriac is tender, about 15 minutes. Add salt and pepper to taste, and serve piping hot. For a more elegant treatment, add a bit of chopped parsley or tarragon to the broth when the celeriac is done, and reduce the liquid over high heat to make a sauce. This dish serves 5 or 6, depending on the size of your celery roots, and is an excellent accompaniment to meat dishes.

Celtuce

Lactuca sativa var. *asparagina*

LIKE ordinary lettuce, celtuce belongs to the daisy family, but this fascinating two-crop variant can be nibbled like lettuce when it's young and munched like celery when it's older. Sometimes called asparagus lettuce or stem lettuce, celtuce has been known to at least some American gardeners since the turn of the century. It's likely, though, that the seeds we're buying today are descended from some a missionary sent home from western China around 1940.

Though it takes as long to mature as some kinds of head lettuce, celtuce produces leaves in the shape of a rosette. Lighter green than regular lettuce, these are comparable in taste and texture to romaine or cos rather than to the more delicate kinds of leaf lettuce. For this reason there seems to be a bit of controversy about the edibility of celtuce leaves. Some gardeners applaud their crispness and resistance to "salad dressing wilt," while the tender-lettuce lovers find "-tuce" barely palatable. In any case, to the joy or despair of those harvesting celtuce leaves, about four weeks after they first appear, a milky sap starts to form. This liquid, which gives the lettuces their botanical name, makes the leaves too bitter even for those who fancy them and ends the first harvest.

At about the same time, more vigorous growth gets under way. As the plant bolts to a seedhead, the stalk that bears the tiny leaves at the top elongates. Reaching heights of up to 5 feet, this celery-like stem remains tasty until the flower buds develop. For the best dining, however, the "cel" part of celtuce should be harvested when the stems are about 1 inch in diameter at the base.

Celtuce seeds can be sown where they are to grow early in the spring, as soon as the soil is workable. Southern gardeners also report growing this vegetable all winter in the greenhouse and setting out plants in February. Since the plant does well in cool weather and can resist a little frost, you can also direct-sow in early August for a fall crop. Like lettuce, celtuce is very tolerant of different soils, but it thrives in well-prepared soil enriched with compost and well-aged manure. Sow your seed about ¼ inch deep in rows 18 inches apart, covering it lightly. When the seedlings are 2 inches high or more, you can start to thin them, using the leaves in salads or moving plants to make another row. Allow about 1½ square feet of space per plant.

About 45 days after planting, you can begin to harvest your leaf crop. As you do, you should side-dress the celtuce with some rotted manure. Perhaps five weeks later, you'll be reaping giant stalks. To harvest them, cut off the stems at ground level and strip off the leaves. Or you can pull each plant out by the roots and cut it off at the crown. The cool-looking celery-like stalks will tempt you to take a bite right there in the garden, but if you do, you're in for a nasty surprise. The milky sap oozing from the cut stem will give you a bitter mouthful, and the bigger the stalk, the worse will be the taste. Many disappointed gardeners have given up at this point, writing off their huge, healthy-looking celtuce plants as a mysterious disaster. You, however, will know better! Before you chomp into that first stalk, peel off the outer skin to get rid of the sap tubes that carry the bitter principle. You'll be left with a soft, cool green core that lies somewhere near cucumber or mild summer squash in flavor. Harvested celtuce stalks will keep very well in the refrigerator, and the leaves store about as well as lettuce.

Because it's disease resistant, easy to grow, and prolific, celtuce is a good bet for folks who can't grow celery and for those who have trouble bringing head and leaf lettuce to full

• Our soil is heavy clay and consequently we have to add a lot of sand and compost in order to get anything to grow. My luck with head lettuce and leaf lettuce has been sad. One of the seed catalogs advertised celtuce and I sent for a package. It said that it had a stalk like celery and a leaf like lettuce. It sounded like a good way to get two vegetables for the price of one.

I planted the seeds in the clay soil by digging a shallow trench and digging in some compost. As soon as the seeds had sprouted and were about 2 inches above the ground I spread about an inch of bark dust as a mulch over the entire row. It wasn't long before we were eating the celtuce leaves.

—*Ralph S. Blois*
Aloha, Oregon

development because of heavy clay soil. Though the central stalk is long in coming and ties up garden space, celtuce does offer two crops for the price of one. Besides, its leaves will give you four times as much vitamin C as head lettuce.

You can enjoy celtuce stalks raw as a finger food served with a dip or sliced into a salad. Their crispness also makes them a natural choice when you're stir-frying vegetables for chow mein, fried rice, or other oriental favorites. For some more good eating, cut some stems crosswise and on the slant into 2-inch lengths, then saute them in oil for about 2 minutes, stirring constantly. Turn the heat down, add a cup of chicken or beef stock and steam for 10 minutes, or until the celtuce is tender but still firm. Salt to taste and garnish with chopped parsley. Celtuce also tastes good au gratin. Just cut the core into 1-inch slices and boil it for several minutes in a little water. Put the drained celtuce in a baking dish and cover with some grated fresh cheese or with a medium white sauce to which you've added grated cheese. Then sprinkle with bread crumbs and cook in a preheated 350°F. oven for 15 minutes or until brown.

Another interesting option is cream of celtuce soup! For this dish, saute 2 cups of diced celtuce in 2 tablespoons of oil until almost tender. Then remove and set aside ½ cup of celtuce, add ½ of a medium-sized onion (diced) to the remaining celtuce, and continue to saute until both vegetables are tender. Next, blend or puree the sauteed celtuce and onions, then start to heat the puree in the top of a double boiler. Add 4 cups of chicken or turkey stock to the puree and keep heating it. Meanwhile, combine ¾ cup skim milk powder with ¾ cup water, using a wire whisk. Stir this into the puree and stock, and leave on the stove just long enough to heat through. When the soup is about ready, add the reserved diced celtuce, ½ teaspoon of kelp powder, and salt to taste. Garnish with parsley and serve to six adventurous eaters!

You can order celtuce seeds from W. Atlee Burpee Co.; William Dam Seeds; Henry Field Seed and Nursery Co.; Gurney Seed and Nursery Co.; Earl May Seed and Nursery Co.; Nichols Garden Nursery; Stokes Seeds, Inc.; and Thompson and Morgan, Inc.

• We are very pleased with our crop of celtuce this year. It is very hardy and has withstood snow and below 30° temperatures several times this spring. It grows faster than lettuce. We've been letting it get approximately 12 inches in diameter before picking the outer large leaves only and leaving the smaller leaves to keep growing, hopefully to use the same way all summer.

—*Barbara Penman*
Wellsboro, Pennsylvania

Chayote

Sechium edule

WHAT has leaves like spinach, tender shoots like asparagus, fruits reminiscent of summer squash and eggplant, nut-like seeds, and tubers that can double for potatoes? The answer is the superplant that rhymes with *coyote*. This fast-growing tropical vine also probably can leap tall buildings at a single bound, since it grows anywhere from 30 to 100 feet in a single season. Chayote was cultivated by the Aztecs and Mayans long before Columbus and other Europeans arrived on the scene. Native to Mexico, Central America, and the West Indies, the plant is now also grown in South America, parts of north Africa, and in subtropical parts of the United States such as southern Florida and Louisiana, Mississippi, southern Oklahoma, and lower California. It can be grown in the greenhouse farther north.

This versatile vegetable is becoming a valued commercial crop and seems to have even more names than it does edible parts. Known in Mexico as chayote, which is a variant of the Mayan word *chayotli*, the plant is called mango squash, christophine, or chocho in South America, vegetable pear in Florida, and mirliton in Louisiana and Mississippi. The French call it *brionne*, while other European gourmets know it as custard marrow or *pepinella*. And in parts of Australia it's dubbed *cho ko*!

Belonging to the same family as cucumbers and gourds, *103*

chayote is a climbing vine with heart-shaped angled or lobed leaves that measure 4 to 6 inches across. It's a perennial in truly tropical areas but an annual in the subtropics and can be quite variable in the size, appearance, texture, and taste of its fruits. Containing a single flat seed from 1 to 2 inches long, chayote fruits range from 3 to 8 inches long, from green to ivory white, and from a few ounces to as much as 2 pounds. They tend to be either smooth and spherical or deeply fissured and resemble an avocado or pear in shape. The chayote may be fiberless and without a noticeable seedcoat or may have a very marked seedcoat from which fibers radiate into the fruit. Resembling white potatoes or yams in taste, the tubers of chayote are even more popular than its fruits in the tropics, where the fleshy roots may weigh up to 25 pounds each.

As the wide diversity in plants hints, there are definite strains of chayotes producing fruits and tubers that vary considerably. So far, however, clear-cut varieties have not been identified. Most home gardeners in the United States seem to work with the seeds of high-quality fruits gotten from the market or from friends, but chayotes grown from seed do not breed true. A surer approach to tasty fruits and roots is spring propagation of small shoots taken by crown division from a good plant that has been kept alive over the winter. This can be done by cutting off the vines, mulching the plant well with hay, covering the hay with a bushel basket, then mounding leaves over it.

In the United States chayote is most successfully grown in the milder regions of the South Atlantic and Gulf Coast states and in southern California. Though the south edge of Georgia is balmy enough for this warmth-loving vine, Atlanta has too short a growing season and fruits grown from seed there failed to mature before frost killed the plants. Interestingly, though, chayote can yield edible fruits and tubers in areas with a hot continental climate in summer if it is started in a greenhouse in mid-March and set out at the very end of May.

If you're tempted to try chayote, remember that this superplant is a very enthusiastic climber requiring full sun, lots of space, and large amounts of water. Locate your plants so you

• We planted the chayote seed in our greenhouse bench and decided this was just the plant we needed to crawl up on strings and sash bars where nothing else grew in the "dead space." The fruit was placed on its side, covered halfway with soil, and that's all it needed to sprout and grow fast.

The vine grows quickly and needs support in the greenhouse. It needs watering almost daily and can use feedings of any fertilizer (fish emulsion is fine, once every 4 weeks). It took the vines about a half year to produce fruit on top and tubers in the soil. We harvested the fruit and made it into a dish resembling squash, but the big bonus was the tubers, to us the best part of the plant.

—Doc and Katy Abraham
Naples, New York

can train them to crawl up and over a fence or on an outbuilding, or perhaps up a tall T-shaped trellis covered with 4-inch wire. Another idea is to put chayote under a pine tree that will allow it some sun and light shade and provide a ready-made trellis. Since one plant can set up to 100 fruits in a season, you might be tempted to stop right there. Well, don't. Chayotes usually have both male and female flowers on each plant, but occasionally the two sexes are on separate plants. This being the situation, you'd best plant at least three chayotes to be sure of getting one that will fruit.

If you're working outdoors from seed, for best results start in spring, after the last frost and after the soil has warmed up. Chayotes prefer a rich, well-drained sandy loam, but will do very well in any good garden soil if you provide drainage, mix in lots of well-rotted manure or compost, and water every four to six days if there's no rainfall or if your soil tends to be light and sandy. To start from seed, you'll need a whole mature fruit that has begun to sprout (you can hasten this process by putting fruits lightly packed in straw in a closet until they germinate). Make a mound of soil, and place the entire fruit on its side with the broad end slanting slightly downward. Then cover this lower end with not more than 2 inches of soil, leaving the stem end exposed. Plant the fruits one to the hill, about 9 to 12 feet apart. In late spring start putting on a mulch of straw, litter, leaves, or lawn clippings to keep moisture in the shallow root system and to discourage weeds. From time to time you also might fertilize with cottonseed and soybean meal to give your lively vines more food to grow on.

Like other cucurbits, chayote is vulnerable to certain diseases and insects. In soils troubled by nematodes, rootknot can convert otherwise perennial chayotes into annuals. Your most practical answer is to build up the humus content of the soil by consistently using compost. You might also try interplanting your chayotes with marigolds of the French or African variety. If cucumber beetles should take a liking to your giant vines, try the time-tested control of heavy mulching. In drastic cases, you might make a spray of a handful of wood ashes and a handful of lime in 2 gallons of water and apply it to both sides of the

leaves. Marigolds, radishes, and nasturtiums can offer more protection through interplanting.

Chayotes fruit best at moderate temperatures. In the most southern states, the plants will fruit in October and November and the vines usually will die back over the winter when night temperatures drop to about 40°F. (If they survive, there will be a light crop of fruit in May.) Farther north, fruiting begins earlier with the coming of cooler weather, and the plants die with the first freeze.

In late spring and early summer, you can harvest some of the chayote's delicate asparagus-like shoots, which have a subtle flavor somewhere between that of celery and cabbage. Just break off the last 10 inches of the thick and tender young stems, leaving on the leaves and tendrils or cooking them separately as greens. Toward the end of summer, try some of the tender young fruits unpeeled and without removing the infant seeds. As the chayotes mature, the peel will toughen and you'll want to remove it and to boil the seed before you eat it. Starch changes to sugar in ripening fruits, so the somewhat older ones will taste sweeter. Left too long on the vine, though, chayote gets tough, and the fruits reach their flavor peak when they're 4 to 6 inches long. If you're growing this vegetable as an annual, you can dig up the tubers at the first frost when the plants die. With a bit of care, however, you can remove and enjoy some tubers from a mature plant well before the end of season. All parts of the chayote are tastiest fresh, but the shoots and fruits will keep pretty well in the refrigerator, and you can probably store the tubers in a cold cellar.

Nutritionally speaking, chayote fruit is low in calories (just 108 per pound) and high in trace elements. Depending on the strain it comes from, it can also be a good source of dietary fiber. And since it's low in sodium and high in potassium, which is usually lacking in low-sodium foods, chayote is a particularly good food for those on salt-restricted diets. The fleshy roots of this plant are also low in calories, and are about as nutritious as potatoes. Indeed, in the markets of Costa Rica these popular tubers sell for two or three times the price of potatoes. But perhaps the best reasons for growing the vegetable pear are its

astonishing productivity and the versatile ways it can be enjoyed.

The succulent shoots make a tasty stand-in for asparagus, so try them boiled for 4 to 7 minutes and served with butter or hollandaise sauce or in a cream sauce on toast. You can also parboil and marinate them for use in salads. Prepare the tubers as you would boiled potatoes or go on to fry them for home fries that taste just like spuds. You can also serve them as low-calorie French fries. In the tropics, the tubers are boiled until tender, then sliced and fried in butter and syrup much the way we candy yams or sweet potatoes.

The somewhat larger chayote fruits can be sliced or diced raw and added to salads or other dishes where they make a fine substitute for cucumbers. Or you can enjoy them parboiled, then marinated for several hours in vinegar, olive oil, and marjoram. To many people, chayote tastes a great deal like summer squash. Sliced ½ inch thick, mature chayote can be used instead of eggplant or summer squash in many ways: as a fried vegetable, in casseroles, or layered with tomatoes and cheese to make chayote Parmesan! Chayote halves also may be stuffed with meat and vegetables or meat and rice, then baked like stuffed eggplant. When sliced, this adaptable food can double for the pasta in lasagna or for the meat in veal dishes, and when diced, it makes a tasty vegetable curry. Try it with a Mornay, cheese, or white sauce, too.

For a pleasant side dish, try ½-inch cubes boiled from 8 to 10 minutes, then added to heated spaghetti sauce. Or you can add cubed chayote to sauteed chopped onion, cook on low heat for a few minutes, then add chopped fresh or canned tomatoes, thyme, salt, and pepper and cook for another 10 minutes. In Brazil, scooped-out balls of chayote are steamed over lightly salted and sweetened water until barely tender. Then they are served hot with butter, freshly ground white pepper, and a dash of cinnamon.

Fully mature chayote fruits may be used in place of potatoes in chowders and vegetable soups. To make savory fritters, shred chayote and add it to a beaten egg, a little milk, and flour seasoned with chopped onion and parsley, then fry.

How To Make
Chayotes Taste Better

Recipes are many, but I never saw advised the way I treat the preparation of chayote. Cut the fruit across its width with a sharp knife and rub the two halves in a circular motion. There will appear between skin and meat a white, sticky liquid that I remove. Now the chayote can be peeled very thin, or when preferred, cooked unpeeled. In that case, just scrub it clean. A chayote treated this way, in my opinion, tastes better.

—Marianka de Nijs
Corona, California

The grated fruit also makes an acceptable variation on carrot or zucchini in breads or cakes. If you have an abundance of very young fruits 2 to 3 inches in diameter, try pickling them. They're good either alone or in mixed pickles, sweet or sour, dilled, or as a chow-chow or relish.

Not content with starring as a main or side dish, chayotes also make a palatable dessert. When boiled and mashed with cloves and lemon juice, the fruit resembles applesauce. It also can be used as a base for tarts and pies with any fruit juice. To enjoy a dessert favored in Guatemala, cook very ripe chayotes in water and salt, then cut in half. Extract the pulp carefully and add to it egg, sugar, cinnamon, nutmeg, almonds, raisins, and cornstarch. (For each large chayote, add one egg, ¼ cup raw sugar, ½ teaspoon each of cinnamon and nutmeg, ¼ cup of mixed raisins and nuts, and 1 tablespoon cornstarch.) Refill the chayote halves with this mixture, putting any extra in custard cups, and bake on a cookie sheet at 350°F. for 40 minutes. Serve with whipped cream or ice cream, either warm or cold.

As if all these possibilities aren't quite enough, the chayote plant also has nonfood uses. The leaves and stems make a nutritious fodder for livestock, and the stems reportedly have commercial potential as the source of strong fibers that can be extracted and spun into cord.

If you're not in a position to get a crown-cutting from a chayote plant or to choose fruits for planting from a market, you can order chayote for planting from the Reuter Seed Co., Inc., 320 N. Carrollton St., New Orleans, La. 70179. Reuter lists this wonder plant as the vegetable pear.

Chicory

Cichorium intybus

CHICORY has been used as a salad plant since ancient times, but was probably not cultivated in its native Europe or Asia until the medieval era. Wild and thick-rooted varieties were introduced as coffee substitutes during the eighteenth century, and produced on a fairly large scale until after World War II. In England, a law was passed in 1832 forbidding the use of chicory to adulterate coffee. But the taste of chicory had become so popular that public outcry caused the law to be repealed in 1840. Today, the many varieties of chicory under cultivation fall into two major categories—those grown for their leaves and those grown for their large roots. Keeping all the varieties separate can be a source of much confusion to the novice grower.

Some of the newer varieties of leaf chicory are grown for their self-blanching leaf heads, which look a great deal like cos lettuce and are used as salad greens. Other leaf varieties, notably WITLOOF, are planted in spring chiefly to develop large, healthy roots which are taken up in fall and forced to produce mild-tasting shoots, called Belgian endive, throughout the winter. The leaves of forcing chicory may be eaten in summer, but if too many are cut, it creates the risk that the roots will be too busy sending up leaves, and will not be able to store up enough energy to force shoots during the winter.

111

Two Ways to Make
Chicory Coffee

I have read in several places that chicory roots made a good coffee substitute. The roots are dried until dark brown, then ground.

My trick, though, is to omit the grinding and drop part of a root with ground coffee into my drip pot. This allows the use of less coffee, and makes a tasty brew.

More important to me, since I am a coffee lover with a delicate stomach, is that the root gathers the oils and seems to gather the acids. At any rate, coffee without chicory can create a burn. But with chicory added, I enjoy coffee with no upset.

—Virginia Stacey
Genoa, Wisconsin

The large-rooted types of chicory, which also go by the name of Italian Dandelion, produce leaves that look rather like smooth dandelion leaves, and large, white taproots. The leaves of this type are used as summer salad greens, and the roots, when roasted and ground, serve as coffee extenders or substitutes.

A third member of the chicory family, Asparagus Chicory, is also called Italian Dandelion on occasion. But this type is grown for its thick leaf stalks, which taste something like asparagus and are served the same way.

A perennial member of the Composite (daisy) family, the familiar blue-flowered chicory so abundant in summer is the wild forbear of the cultivated types. Wild chicory grows rampant in pastures, fields, and waste places throughout the United States, but especially in the eastern and central states. Its leaves can be picked in early spring for salad greens or a potherb, and its roots may be gathered in summer or fall for roasting.

Leaf chicory is the most commonly cultivated type in this country. Some varieties, such as SUGAR HAT, are grown strictly as summer or fall greens. If you want to grow chicory with the aim of forcing Belgian endive during the next winter, your purpose will best be served by choosing a pure forcing strain, like the popular Witloof (the name means "white leaf" in Flemish). A red-leafed variety, ROUGE DE VERONE, forms small, self-blanching heads, and is suitable for forcing, as well. All the leaf varieties have broad leaves and wide midribs, and when blanched, form compact heads similar to romaine lettuce but shorter. This is the kind of chicory familiar to most gourmets, and most often confused with endive and escarole. It is also called French endive, or succory.

Large-rooted varieties of chicory, the type used for spring greens and coffee substitutes, include MAGDEBURG, BRUNSWICK, and PALINGHOP. These varieties are perennial crops, and may be grown from seed in permanent beds.

Non-forcing chicory is usually seeded in June or July, while most of the forcing types are planted earlier, in May or June. In cooler regions, chicory for forcing should be started indoors in

May. In warmer areas where the growing season is five to six months long, leaf chicory should not be sown too early, or it may go directly to seed. Instead, plant the crop in mid-June in order to encourage the formation of large storage roots suitable for forcing.

For all varieties, sow seed thinly in rows about 18 inches apart. The seedlings, when established, are thinned to stand about 6 inches apart. Any light, fertile soil suitable for root crops such as beets, carrots, and parsnips will be satisfactory for chicory. Cultivation and care are the same as for parsnips. The plants appreciate a good layer of mulch and adequate water, although they can tolerate drought. Keep the rows free of weeds. Once established, the plants are quite hardy.

As the plants grow, they put out dandelion-like leaves of a medium shade of green. Leaf varieties will go on to form heads. Leaf chicory that is grown for greens is sometimes self-blanching. Otherwise, the heads need to be tied in order to blanch the inner leaves and reduce bitterness. When the heads mature in fall, they are cut.

The curly-leafed tops of forcing chicory will begin to wilt and fade in late October or November. After the first hard frost, but before the first long freezes set in, dig or pull up the roots and trim off the tops, leaving a 2-inch stub above the crown. The roots can then be placed horizontally in sand or peat moss and stored in a cool (below 40°F.) place, where they will keep for several months. Harvested taproots may range in size from 5 to 10 inches in length and from ½ to 2 inches in diameter. The best roots for forcing are about 6 or 8 inches long. Larger than that they tend to form tough, divided sprouts.

Although root chicory will tolerate freezing, and will come back in spring, these varieties should be harvested in October in areas where the winters are extremely cold or for possible winter use. Dig the roots, remove the tops, and store the roots in the same manner as roots that are stored for later forcing. The roots can be put back out in the garden the following spring.

Roots for forcing should be kept in storage for a month

• The chicory roots are ready for roasting in the fall. They are washed and cut into sections not more than ¼ inch thick and roasted until they are dark brown all the way through. Pound the chicory with a hammer, crushing them, or grind the roots in a coffee mill. Chicory prepared like this can be made into a beverage in the percolator or by any method used for coffee. Use about three quarters as much chicory as you would coffee to make a brew of comparable strength.

—Adele Laux
Fauquier, British Columbia
Canada

Here is an old-fashioned way to serve chicory, from *The Book of Rarer Vegetables* (published by John Lane Co. in 1906):

The green leaves of Chicory may be cooked in spring by scalding them for five or six minutes in boiling salted water, draining through a colander, throwing the leaves into cold water, again draining, and then cutting up the leaves. The chopped leaves should next be placed in a saucepan containing (for each pound of fresh Chicory leaves) half an ounce of flour, a quarter of an ounce of butter, and a little pepper and salt, which has been heated over the fire for three or four minutes. Stir all together over a gentle fire for five minutes, and then add a teacupful of milk or broth, and stir over the fire till nearly dry. Take the pan from the fire and add about an ounce of butter, stir, and serve on a hot dish. Always use plenty of water in blanching the leaves.

before the process is begun. To prepare roots for forcing, trim off the slender tips so the roots to be forced are of an even length. Set the roots in a trench with the crowns about even with the soil surface. They should be placed closely together to encourage straight, firm shoots. Any box or flower pot can be used for forcing chicory, provided it is filled with damp sand, loam, or a mixture of sand and peat moss. One excellent method involves filling a large orange crate with 6 inches of composted manure, 6 inches of fine loam, the chicory roots, and then dry peat or sand to cover the crowns. This top covering excludes light and prevents leaves from spreading. Before it is applied, the roots should be thoroughly watered. One or two waterings may be necessary later, but the soil above the roots should never be soaked. Decaying matter, manure, or even dampness in this layer could cause the heads to rot.

The forcing box or pot is then placed in a completely darkened, warm place. A cellar, cupboard, garden shed, or barn is suitable for forcing, so long as slight warmth is supplied. Forcing can also be done under greenhouse benches. If a layer of uncomposted manure is used, the box can be placed out of doors, for decaying compost will generate some heat and protect the developing shoots from frost. The cold frame is a good place to force chicory if the temperature can be kept between 50 and 60°F.

Examine the box in two or three weeks and, if the heads or sprouts have poked through the soil surface, they are ready to use. The best Belgian endives are 5 to 8 inches long and weigh 3 or 4 ounces. If the heads, or chicons, as they are also sometimes called, are too small or are not wanted by the time the tops emerge, then add another inch of cover soil.

To harvest Belgian endive, break or twist the heads about ½ inch above the crown without injuring the growing point. Clean them with a soft cloth from the base to the tip and use as soon as possible. Tamp down the peat, sand, or loam after the first harvest, and within several weeks, a second crop will push through the surface.

The chicon, formed from the closed rosette of leaves, is oblong and whitish, except for the light green leaf tips. Blanching and forcing the shoots in this way removes most of the bitter quality present in the green leaves, and makes for a very tender, delicately flavored vegetable. Belgian endive is greatly esteemed by American and European gourmets, who enjoy it simply steamed or braised. The shoots are mild enough to serve raw with a vinaigrette sauce or *a la grecque,* in salads, or boiled and accompanied by your favorite cream sauce. They are especially delectable when served in a light cream sauce flavored with grated Parmesan and just a touch of dry mustard. They are also delicious additions to soups and stews.

To harvest the taproot of root chicory for use as a coffee substitute or additive, gather the perennial roots after the flowering season has ended. Scrub them well, lay them on a screen or wire rack, and let them dry for several weeks. Chop the dried roots and roast them until crisp, then grind as for coffee.

The leaves of large-rooted and wild chicory possess a bitter quality that many people appreciate in a tossed green salad or Italian-style salads. The leaves can also be cooked like spinach, and a change of cooking water will remove some of the bitterness. Stir-frying also takes away the bitterness, and stir-fried chicory is quite tasty.

The roots of both the wild and cultivated varieties of chicory can also be used as vegetables. They can be handled in much the same way as parsnips—braised, steamed, baked, or boiled and served with your favorite sauce.

Witloof chicory is available from most seed suppliers. This and other leaf varieties can be had from W. Atlee Burpee Co.; Comstock, Ferre and Co.; William Dam Seeds (who carry 6 varieties); J.A. Demonchaux Co.; DeGiorgi Co.; Joseph Harris Co., Inc.; Charles C. Hart Seed Co.; R.H. Shumway Seedsman; Stokes Seeds, Inc., and Thompson and Morgan, Inc.

Suppliers of large-rooted varieties to use for coffee include Comstock, Ferre and Co.; William Dam; Le Jardin du Gourmet; Hart; and Stokes.

The common blue-flowered strain is available from Meadowbrook Herb Garden.

Collards

Brassica oleracea, var. *acephala*

DIFFERING from kale only in the shape of their leaves collards, too, are a nonheading primitive form of cabbage. Believed little changed today from its wild prehistoric ancestors, the collard was grown in ancient Greece and Rome where it was not distinguished from kale. This enduring vegetable was probably carried to Britain and France by the peripatetic Romans or by earlier Celts. The first recorded reference to these easy-to-grow greens in the New World dates from 1669, and they've been a favorite in gardens throughout the South ever since. Surely they deserve to be better known in other parts of the country.

Though collards have never achieved the commercial success enjoyed by more evolved relatives like cauliflower, broccoli, and head cabbage, they are the most dependable member of the Brassica family, yielding bushels of mineral- and vitamin-rich greens off a short row. As another bonus, collards withstand a greater range of temperature than any other vegetable grown in the South. Undaunted by temperatures as low as 15°F., they actually improve in flavor after frost, which apparently boosts their sugar content. On the other hand, collards will bolt to seed at the first sign of approaching spring.

In the coastal South, collards are planted from July to

November for winter use, and farther north they can also be enjoyed as a fall crop if the plants are set out six to eight weeks before the first expected freeze. Timing is important, though, for collards planted too early are vulnerable to late summer cabbageworms, while those put in too late won't be able to mature before cold weather puts them on "hold." In the inland South and other areas with mild enough winters, seeds are sown in January for an early spring crop.

A biennial, the collard puts forth a bluish green rosette of broad, coarse leaves that are rather strong in flavor. The best-known variety is the GEORGIA, a rugged plant that often exceeds 4 feet in height and has a huge mass of loose foliage toward its top.

To grow collards, start your seeds in a cold frame or greenhouse, then transplant your seedlings to the garden when they're 6 to 8 inches high, spacing them 18 to 24 inches apart in rows 3 to 3½ feet apart. Or you can sow seed in a well-prepared permanent row, or broadcast it in a permanent bed, thinning and transplanting when your plants are 6 inches tall. (It's good practice to leave 3 feet between heading collards.)

This unfussy green grows well in all kinds of soil, but it will reward you most handsomely in sandy loam or silt loam enriched with aged manure. According to Cornell University's College of Agriculture, you would also do well to interplant collards with tomatoes, since this decreases trouble from the flea beetle, a prime pest where collards are concerned. If cabbageworms should start to punch holes in your collards, hand-pick them off the underside of the leaves in early morning. Then dust with wood ashes or pickling lime. If your greens are badly infested, you could try *bacillus thuringiensis*, a biological spray sold as Thuricide. Don't be in too much of a hurry, though: the first freeze will put an end to the loopers' lunching. If you should be troubled by aphids on a spring crop of collards, wash them off with a fine spray from the garden hose.

Your greens will be tastiest when they're young, fresh, and bright green, with small midribs. For best results, don't harvest the first six leaves or the central growing tip and gather the lower leaves weekly when they're as big as your hand. Collards make good eating, though, until they're about 10 inches long, so if you like you can harvest the whole rosette at once when the first leaves have reached that size. In deciding how to harvest, you might keep in mind that heading collards won't head if you keep breaking off the bottom leaves.

As you might guess, extra collard greens will store beautifully in the freezer. First trim off the tough stems and discard any injured leaves. Then just steam them for ½ minute at 15 pounds pressure in a pressure cooker or for 4 minutes in a pot; cool, pack, and freeze.

In addition to being widely adapted and among the very easiest of all vegetables to grow, collards are a nutritional winner, containing almost eight times as much vitamin A as cabbage and over twice as much as broccoli. A 1-cup serving also provides as much vitamin C as an orange and nearly as much calcium as ¼ pint of milk. All of this helps explain why early nutritionists discovered—to their surprise—that the health of low-income people in the rural South was quite good.

To prepare collards for cooking, wash them in cool water to remove any dirt and grit, then cut off and discard the tough stems. Those who relish country cooking serve collards in a rich variety of traditional dishes, but all who love these greens agree that fast initial cooking is necessary to keep them from turning bitter. One classic approach is to heat 1 tablespoon of fatback bacon, ham drippings, butter, or oil, and stir in 6 cups of cut-up collards until the greens are coated. Then add water and salt to taste, and bring rapidly to a boil. Cook over high heat until the greens are thoroughly wilted, then lower the heat and simmer until done. Collards prepared this way taste scrumptious with pan-fried fish and hush puppies.

If you want to cook collards in a pressure cooker, start by putting a ham bone or diced ham, salt pork, or bacon in a pot. Add a little water and cook slowly for an hour. Then pack washed, chopped collards into a pressure cooker. Add ½ cup of the "pot likker" from the ham, and steam with the vent open for half a minute, then cook at 15 pounds pressure for 4 minutes. Pour in the rest of the ham and the pot likker and enjoy!

• I plant the Georgia varieties of collards. Probably they would appreciate full sun, but in my small garden I plant them in an area partially shaded by the asparagus. When the asparagus dies down after the first freeze, the collards get more sun. I plant collards (in fact all the brassicas) in the fall garden, trying to get them in the ground the first part of August. I spade in generous amounts of manure. Also compost, if I have any. Then I sprinkle on cottonseed meal and fireplace ashes, and rake well. The rows are spaced 2½ to 3 feet apart, and the seeds dropped every 4 inches or so.

Since loam is something I dream about and clay gumbo is what I have, I cover the seeds with a mix of sand and peat moss. Usually Houston has a superabundance of rain, but if it should be dry, I sprinkle lightly every couple of days until sprouted. Every seed seems to germinate.

—Virgie F. Shockley
Houston, Texas

You can buy your collard seeds from Burgess Seed and Plant Co.; W. Atlee Burpee Co. (Georgia variety); D.V. Burrell Seed Growers Co.; Comstock, Ferre and Co.; William Dam Seeds; DeGiorgi Co.; Joseph Harris Co., Inc.; Charles C. Hart Seed Co.; H.G. Hastings Co.; J.L. Hudson, Seedsman; Jackson and Perkins Co.; Le Jardin du Gourmet; Earl May Seed and Nursery Co.; Mellinger's, Inc.; L.L. Olds Seed Co.; Geo. W. Park Seed Co.; Redwood City Seed Co., R.H. Shumway Seedsman; Stokes Seeds, Inc., and Otis S. Twilley Seed Co.

• For a real great Southern-Style meal, add a few chunks of ham and plenty of water to the greens as they simmer. Meanwhile, bake a big panful of cornbread. Fill a plate with greens, and use cornbread to sop the juice. To complete the protein, serve black-eyed peas alongside.

—Nancy Pierson Farris
Estill, South Carolina

Comfrey

Symphytum officinale, spp.

LIKE many people, James Jankowiak used to think of comfrey as just another weed to feed the chickens. But now he wonders why everybody doesn't eat it, and remembers well his introduction to the finer aspects of this useful plant. "I recall an aged black lady who visited our northern California ranch some years ago," he says. "She took a walk through the garden and discovered the comfrey patch. She was genuinely enthused to see the plant, which I grew as a chicken feed, and asked if she could have some greens. 'Honey,' she said, 'I've been looking for comfrey for years, and haven't been able to find it anywhere. Do you think I could have a few starts?'

" 'Of course,' was my instant reply, 'but what's all the excitement about comfrey?'

" 'You just don't find better greens, honey,' came the answer. 'They're *so* good with a pat of butter melted over them. And they're so good for you—lots of vitamins and minerals. Why, don't you know comfrey will heal a cut better than anything?' "

This ancient plant, with its superior nutritional value and its medicinal qualities, has been highly regarded by rural folk in Europe and the United States for generations. It has been called boneknit, knitbone, or healing herb, and it is still known by those names today. Seventeenth-century British herbals

123

recommended mixing the juice extracted from mashed comfrey roots with wine, and drinking the beverage for internal hemorrhages, or using the mashed root in a poultice on open wounds. Today we know that the curative powers of comfrey (which do indeed exist) are due to a substance called allantoin, which is found in the leaves and roots of the plant. Comfrey can be used for so many different things that you may wonder why it is so seldom seen in catalogs.

Comfrey seed is hard to get for the simple reason that not much of it is produced by the growing plants. The bell-shaped flowers of comfrey have a sort of false bottom that discourages many kinds of bees from getting in to steal nectar. The flowers can only be pollinated by a certain type of bee, which can get through to the nectar. Thus, a lot of flowers never get pollinated, and are never able to set seed. The way to establish a bed of comfrey is to get root cuttings (which are taken from the taproot) or crowns (which are cuttings taken from the roots which grow closest to the soil surface). When you plant your comfrey bed, put it in a place where it can stay permanently. The plant is a perennial, and once established, comfrey is as tenacious as a weed. Like crabgrass, it will come back from even a small piece left in the ground. Unlike crabgrass, however, it will not spread from the area where you plant it. If you're the kind of person who likes to change his or her garden plan every year, put your comfrey in a corner or in a separate bed. It may take you several years to get all of it out if you decide to move it.

The comfrey plant itself is quite pretty, and makes an interesting specimen in border plantings calling for plants of medium height (about 2 feet) and coarse appearance. It has lance-shaped leaves, a squarish stem, and will get quite bushy if encouraged with a sunny location, ample space, compost, and abundant water. Once it starts flowering it continues to do so until frost, and sometimes beyond. The flowers hang down and are usually light purple and shaped like the fingers of a glove.

Comfrey is sure to be a success in your garden. It will grow almost anywhere. It likes a moderately sunny spot, but will also do well in direct sun or shade. It can survive prolonged periods of drought. It will grow in poor soil, even in hard clays. If you take your first starts of comfrey from the wild, or from a friend's patch, the best procedure is to dig the plants and divide the crowns, so that each is left with a few leaves and a piece of root attached. One comfrey grower advises placing the cuttings in blocks measuring 12 inches by 12 inches, in an area that will not hamper normal garden operations. Cover each start to the depth of the crown, but don't allow dirt to cover the developing rosettes of leaves. This procedure is not as critical as it is with strawberries, which often rot if the crown depth is not just right. Comfrey, when established, has a deep, wide-ranging root system that will mine the subsoil for minerals. Thus, it is an ideal compost plant as well as a nutritious food.

Boneknit will thrive with very little care if you assist it in the beginning by keeping the patch free of weeds. Once it gets going, comfrey will choke all but the most pernicious weeds. Despite its ability to tolerate adverse conditions, comfrey needs lots of water and food for peak production. When your patch has reached the size you want, after dividing the crowns each year for several years, the only maintenance you'll have to do is loosen the ground in early spring, work in a few inches of compost, and cover the bed with a thick mulch. If you happen to live in a coastal area, a seaweed mulch is the best to use, as it contains a wealth of minerals that comfrey is able to convert into palatable food for both humans and animals.

Despite the claims you may have heard, comfrey is not a miracle plant that you can simply cut and cut without putting anything back into the soil. It is not a nitrogen-fixer like clover or alfalfa. You can get cuttings from the same plant every couple of weeks for fresh use or hay, but if you never return any compost or mulches to the soil, you'll soon exhaust it, as happens with any other crop.

When comfrey is grown primarily as a fresh vegetable, it needs to be cut back several times a year, whether or not you eat all the leaves. By mid-May, the growth should be pruned back to a few inches above the ground because the plant loses nutrient value if allowed to stand and bloom. You will probably need to cut back the plants two or three more times before the end of autumn. If you don't you may find that the weight of the large, uncut leaves and stems will cause the massive plants to collapse. Cut back the plants for the final time before winter arrives in earnest. New leaves will appear in spring.

Southern gardeners generally find that their comfrey grows slowly during the heat of summer, but takes off in fall and continues growing throughout the winter.

If you are using comfrey for hay, you may find that it takes a little longer to cure than grass or clover. That's because comfrey leaves are thicker, and have a wider surface area.

Comfrey is not bothered by insects, and the only disease to which it is subject, comfrey rust, is extremely rare.

A comfrey crop really begins to hit its peak during the third year of growth. The plants will keep on coming up year after year, but around the tenth year, the crop will begin to decline. Then it's time to dig out the roots and start again with fresh ones. Getting rid of the roots is not an easy chore, and it may require some digging two or three years in a row. But it can be done, and here's how. In fall, dig up all the roots you can find. Try to get all the little pieces cut off by the shovel as well. When you've finished digging, leave the ground rough. Comfrey roots are very hardy, and can survive temperatures of 0°F. and even below. But if the winter is a hard one, the uneven ground surface may increase exposure, and the roots you missed in your fall digging may freeze out. If you do find some stragglers coming up the next year, dig up the roots as soon as you see the plants coming up. Whatever roots are left will eventually quit in exhaustion if no leaves are allowed to grow to nourish them.

Disposing of the comfrey harvest each time you cut back the plants is no problem. The young tender leaves (which can be picked at any time), are best for eating. Their taste is pleasant, although not as exciting as that of some other greens, such as sorrel, for example.

Still, some people find the taste superior to that of spinach. Comfrey has a faint taste of cucumber, but the flavor is less pronounced than that of its cousin borage. For best flavor and tenderness, the leaves should be eaten when young, preferably not more than 6 inches long. As they grow larger, the leaves develop stringy veins and bristly hairs that make them tough. Comfrey can be chopped and eaten raw in salads, or cooked like spinach. The leaves can be sauteed with other vegetables, made into a unique cream soup, added to other soups, and served with a white sauce or au gratin topping. The leaves can also be blended with fruit juice and frozen into ice pops. Jane Wang Holm, a California gardener, reports great success using comfrey leaves to make tortellini dough. Using a blender, she pureed some leaves with plain water to use as the liquid in making the dough. When served, the red tomato sauce over the green tortellinis was quite an impressive sight.

Comfrey leaves can also be blanched and frozen for winter use. The roots of comfrey are good to eat, too. Some people cut up the peeled roots to use in vegetable soups.

When the leaves grow too large and tough to use as fresh greens, there are still lots of uses for them. One San Francisco family uses them for summertime picnic plates to hold tempura or salad. It's fun for both the kids and for non-gardening visitors.

Animals really take to comfrey leaves in their fodder, too. Chickens that eat fresh comfrey as part of their daily diets are said to lay more eggs with bigger, deeper yellow yolks, and seem to have fewer health problems. Rabbits also relish comfrey, and also stay in peak condition when it's part of their regular diet. The comfrey won't cause diarrhea in rabbits the way some other greens will. It is also said that comfrey-fed cows give more milk. Any plant that can produce such good results in animals has got to be rich in nutrients that can benefit people, too. Comfrey has lots of vitamins A and C, and contains large amounts of calcium, phosphorus, and potassium. It has about as much protein as alfalfa; some growers claim it has even more protein, and is more productive besides. What's more surprising, it has been widely argued (though not yet proven) that comfrey contains vitamin B_{12}, a vitamin almost never found in plants. If comfrey does indeed contain B_{12}, it will be invaluable to vegans (vegetarians who eat no animal products whatsoever), who must otherwise take supplements to get this vitamin in sufficient quantity.

As we mentioned earlier, comfrey contains a substance called allantoin, whose healing qualities are well known. Allantoin is useful in treating cuts, bruises, and even ulcers. A poultice of fresh leaves or the powdered root can be used externally to treat excessive bleeding, open wounds, bruises and sprains. Dorothy Onek of Kansas uses a tea made from fresh leaves to soothe stomach ailments. It may also be given for sore throats, sore gums, or ulcers. Sweeten the tea with a little honey. A decoction of comfrey root, mullein and horehound was used by pioneers in the Midwest to make a cough syrup. If allowed to cool, the tea (unsweetened) can also be used for an eye wash.

Dried comfrey leaves make a tea that's pleasant just as a beverage. If after all this, you still find you have some leftover comfrey, chop it up for mulch, or add it to the compost pile. Its high mineral content makes it a valuable addition to garden soil.

Comfrey is available from: Casa Yerba; William Dam Seeds; Gurney Seed and Nursery Co.; Hemlock Hill Herb Farm; Nichols Garden Nursery; Redwood City Seed Co.; and Well-Sweep Herb Farm.

Here's a favorite comfrey recipe of the Sisters at Our Lady of the Rosary monastery in Summit, New Jersey:

Take 12 tender leaves, wash and drain overnight. Wrap each leaf around bread crumbs and one slice of cheese, fastening with a toothpick. Beat 3 eggs and 1 tablespoon grated cheese together, adding garlic, parsley, salt, and pepper to taste. Dip the stuffed comfrey in the egg batter and fry in oil until golden.

Coriander

Coriandrum sativum

IF parsley is an essential part of the vegetable garden, then coriander cannot be overlooked, either. The tiny seeds of coriander are well known as a seasoning, and can be found in any grocer's spice rack. But the fresh green leaves possess an aromatic quality prized by cooks who are familiar with it, and the only way to get fresh coriander in most parts of the United States is to grow your own. For vegetable aficionados and vegetarians, coriander adds a unique zest to many dishes that relieves the blandness even dyed-in-the-wool vegetable lovers are prone to complain of in weaker moments. For instance, new carrots are one of spring's great treats, served piping hot with a whisper of butter and liberal sprinklings of parsley. Later in the season, less-welcome carrots past their prime can be equally as appetizing when garnished with chopped coriander.

Distinctive Mexican dishes are created by cooks who will not tell you that fresh green *cilantro* (the Spanish word for coriander) is their secret ingredient. Likewise, there are Chinese dishes that cannot leave the kitchen in all honesty without *Yuen tsai* (coriander in Cantonese).

Chinese or Mexican parsley, as coriander is also known, originated in the southern European area, but is today cultivated all over the world. The plant forms flat, fan-shaped compound leaves with jagged edges, very similar in appearance to those of flat-leaf, or Italian, parsley.

• Having lived in old Mexico for more than two years, I became fond of some of the really fine Mexican dishes such as *posole*, *birria*, *barbacoa de cabrito*, etc. At every little taco stand in Chapala and Ajijic in the state of Jalisco one could buy *tacos de cabeza* which were delectable. I noticed that on most of these foods, the Mexicans sprinkled chopped greens which looked like parsley but were far more pungent.

. . . When I came back to the United States I missed some of the delectable Mexican dishes and decided to see if I could find out how to make the garnishes myself. I learned that the green herb the Mexicans call *cilantro* is simply the foliage of coriander. So I got a package of seed, put it in a small bed and spent the rest of the summer trying to harvest it as fast as it grew. I froze a good deal of it and finally gave up in dismay when the stuff got ahead of me and began to flower and make seed. I did gather some of the seed and this year decided a half dozen plants would be all I could handle. They are doing well in my garden.

—*Lorene C. Nelken*
Clinton, Iowa

Coriander is easy to grow, and wants a cool temperature to do well. It resists light frosts very well; consequently it can be grown late in the year. In regions where the winters are mild, such as in parts of California and southern areas, coriander winters over nicely. In any climate, coriander is best grown as an early spring or fall crop if it is to be used for its leaves. The most important point to remember when planning planting dates is that hot weather causes coriander to bolt soon after coming to maturity. You will need to replant every three or four weeks to maintain a constant supply of fresh leaves throughout the summer. The plant matures in about 60 days.

Sow seed about ½ inch deep and 3 to 4 inches apart in rich garden loam. A well-worked soil is important to the success of the crop, for the seed may fail to germinate in hard clay. For a small family, about 2 feet of row will provide an abundant harvest of coriander. Keep the soil well watered for two weeks. Some gardeners recommend placing a board over the short row to keep the earth moist, taking care to remove it in plenty of time for the seeds to come up unhindered.

When your coriander comes up, you'll find the first leaves look very much like parsley. Coriander will respond to fertility, but it will manage to give you a return in poor soil as well. Coriander grown in ordinary, unfertilized soil is generally considered to be more pungent than that grown in extremely rich earth.

Harvest of the individual leaves can begin as soon as the mature leaves appear, and can continue until the plant dies back. Fresh coriander does not keep too well once it's picked, but for long-term storage, the leaves can be chopped and frozen like parsley, without blanching.

You will recognize the distinctive flavor and aroma of coriander in such characteristic dishes as Mexican *salsa,* Chinese smoked duck with plum sauce, or Chinese-style barbecued beef. Coriander also makes an excellent seasoning for many pork, fish, and egg dishes. Try it in scrambled eggs or an omelet for a really lively breakfast.

Just about every company that handles herbs sells coriander. Suppliers include Burgess Seed and Plant Co.; W. Atlee

Burpee Co.; Casa Yerba; Comstock, Ferre and Co.; William Dam Seeds; DeGiorgi Co.; Charles C. Hart Seed Co., H.G. Hastings Co.; J.L. Hudson, Seedsman; Jackson and Perkins Co.; Le Jardin du Gourmet; Johnny's Selected Seeds; Meadowbrook Herb Garden; Mellinger's Inc.; Nichols Garden Nursery; L.L. Olds Seed Co.; Geo. W. Park Co.; Redwood City Seed Co.; R.H. Shumway Seedsman; Sunrise Enterprises, and Thompson and Morgan, Inc. Well-Sweep Herb Farm can supply either seed or plants.

• Since discovering myself the pleasures of fresh coriander I cannot do without it. Some dishes are just not prepared without it. . . . Mexican friends, on visiting my garden, have hunkered down with me near the *cilantro* and eaten whole handfuls of the stuff. . . Some doleful Vietnamese students living in our town came to life and started chittering like squirrels when presented with a bunch of coriander. They were delighted that they could find one of the special tastes of their home in this country.

—John Meeker
Gilroy, California

Corn Salad

Valerianella locusta, var. *olitoria*

THROUGH much of the midwestern and southern part of our country, growing lettuce is somewhat a problem because of the high temperatures of early spring. Lettuce, a garden essential for nearly everybody, is easy, foolproof, and seldom troubled by disease or insect pests. But to grow to perfection, it does require a cool growing season. Gardeners in northern states can raise just about any kind of lettuce they wish, but gourmets in warmer sections of the country must usually content themselves with the familiar leaf lettuce, and the softer-heading varieties.

Corn salad promises to change all this, and add some variety to salad bowls in the South, and in the rest of the country as well. It isn't really new. Corn salad was known and sometimes grown by our gardening grandparents, although it never achieved great popularity. But its versatility will make it a welcome addition to your home garden. Corn salad can be grown throughout the United States—as a spring crop in the North, and as a winter crop in the South.

This subtle-tasting green goes by several names. The most common one, corn salad, has nothing at all to do with American sweet corn; rather, it comes from the presence of the plant as a troublesome weed in the grain fields of Europe. It used to be customary in Europe to refer to any grain as "corn"; thus, even

135

though the plant grew in the wheat and rye fields, it came to be called corn salad. It is also known as lamb's lettuce, and two interesting stories are told to account for that name. One theory has it that sheep are particularly fond of this plant, and will seek it out in preference to other more available weeds. The other explanation is that the vegetable is at its best when harvested at the end of winter, when the lambing season begins. The French word for lamb's lettuce is *maches,* a name which has passed into English usage perhaps on the strength of Parisians' reputations as gourmets. Great quantities of leaves pass through the Paris produce markets from spring to fall, to the delight of urbane Frenchmen. Yet another name for this well-traveled plant is fetticus, an epithet universally deemed unappealing by gardeners, some of whom even ascribe the vegetable's lack of wide popularity to the sound of its name.

To the more scientific-minded among us, this plant is a member of the valerian family, now naturalized in many places throughout the United States, where it grows either under cultivation or as a weed. A related species, *v. eriocarpa,* or Italian lamb's lettuce, is cultivated in Mediterranean areas, but has not found a foothold in the western hemisphere because of its unreliability in northern climates. Except for the Italian variety, which has hairy leaves, all corn salad looks pretty much the same. Spoon-shaped, rounded leaves, attached to the near-ground-level stem, grow about a foot long. Sometimes the leaves grow in rosettes, and some types form loose heads. For the most part, though, corn salad resembles deer tongue leaf lettuce, and requires about the same care.

Corn salad is generally easy to grow. It will grow in nearly every type of soil, and really thrives in soil that is well-enriched with a high-nitrogen compost or manure. In most parts of the country, corn salad may be sown twice a year. The first sowing is made in early spring, as soon as the ground is workable. Lamb's lettuce likes cool weather, and can stand some frost. Early sowing allows the crop to mature in cool temperatures it likes so well. In most parts of the country, a second sowing may be made in August or September, so the leaves will mature in autumn. The big advantage to corn salad over other types of greens is its ability to withstand frost. In fact, one California gardener has found that her corn salad crop has survived freezes hard enough to put a half-inch layer of ice on the water trough. She reports that, "when on occasion I would go out in the morning to pick a salad the leaves would be frozen crisp and translucent. But as soon as the sun came to melt the ice, they perked right back up and suffered no harm."

In areas where the ground freezes solid, a cold frame can allow the plants to grow through harsh weather. A heavy layer of mulch sometimes allows plants in northern areas to survive the winter and begin rapid growth early the following spring. Another way to get an early spring crop is to sow seed broadcast in late fall in a rich soil, cover with a light mulch, and in spring the plants will germinate quite early to provide some of the first fresh garden vegetables. In mild climates the plants will stand in the rows unprotected all winter. In areas where the winters are quite severe, a single planting in early spring is probably the best practice.

Sow the seed thickly in furrows ½ inch deep, in rows about 12 inches apart. Germination is not always the best, and sowing heavily will help insure a good crop. After the seedlings are up, thin them to stand 4 inches apart.

Some experts recommend growing lamb's lettuce continuously from spring until September. In areas where summers are cool, this method works well. However, gardeners in regions where summer brings hot weather have found that while the leaves remain in good condition, growth slows considerably. Dale Hilden of Texas, though, has found a way to germinate and grow lamb's lettuce even in blistering summer heat. Here's how he does it:

> Our garden rows run north and south, and so the east side of the sweet corn rows has afternoon shade from noon on—which provides just the right protection and conditions to encourage growing fetticus. With the first cool days of September, we plant fetticus for the fall garden and find that again it is as hardy as turnips and supplies a delicious addition to salads and greens until killed by a hard freeze.

Under optimum conditions, corn salad will mature in 50 to

• Here's one gardener's first-time experience with a crop of lamb's lettuce that matured in summer:

"I planted my seeds in the spring. They germinated well, forming tiny little rosettes close to the ground. I waited for it to get big enough to pick when suddenly the tiny plants were covered with longish stems and clusters of tiny blue flowers. And that was the end of their life span. What a waste, I thought. I dug them under and planted carrots."

"Then, in early fall, as the carrots were almost all pulled, appeared a thick layer of tiny seedlings, dense as grass. Since I know all my weeds I observed these newcomers closely and in no time at all they bushed into lovely mounds of dense leaves—the whole cluster about the size of Bibb lettuce. My enquiring nibble was a delightful surprise—taste just like rose petals. As I pulled carrots the corn salad emerged to take their place until the whole patch was a solid carpet. All through the winter I pulled out the plants for wonderful salads and more quickly filled their place."

—*Joanne Stephens*
Fort Bragg, California

60 days, though you can pick individual leaves much earlier than that. Optimum conditions means full sunlight, ample water (re-watering whenever the top half-inch of soil dries out), and nitrogen-rich soil. If you are growing a leafy variety, leaves may be picked one at a time after they are a few inches long, or you can wait until the plant matures and harvest the entire head at once. Some people prefer to blanch the heads as they mature, but this isn't really necessary, and is strictly a matter of choice.

Corn salad has a unique flavor that has been described as mild to the point of blandness, with just a hint of nuttiness. One gardener compared the taste to roasted, unsalted soybeans; another said the leaves when raw taste like rose petals, and when cooked, like garden peas. Most people find that the texture of corn salad leaves is unlike any other salad green they've ever eaten. The leaves are not crisp like lettuce, not succulent like New Zealand spinach, but are instead so delicate they seem almost to melt in your mouth. One California gardener says everyone to whom he's ever given a sample of lamb's lettuce has commented, "It's sure different," and all have liked it.

Corn salad has a variety of uses in the kitchen. It is a welcome addition to the salad bowl, where its mild flavor nicely complements more strongly flavored greens like chicory and cress, and crisp root vegetables such as carrots and radishes. Lamb's lettuce is also valued as a potherb, to be cooked quickly and seasoned like spinach. Europeans have traditionally served corn salad in connection with beets, and to the American palate, it combines pleasantly with other early-season greens, such as sorrel.

Corn salad, lamb's lettuce, maches, or fetticus is available from a number of suppliers, including Casa Yerba; Comstock, Ferre and Co.; William Dam Seeds; J.A. Demonchaux Co. (who list three varieties); DeGiorgi Co.; Gurney Seed and Nursery Co.; Charles C. Hart Seed Co.; J.L. Hudson, Seedsman; Le Jardin du Gourmet (who list four varieties); Nichols Garden Nursery (who call it maches); Redwood City Seed Co.; R.H. Shumway Seedsman; Stokes Seeds, Inc., and Thompson and Morgan, Inc.

Cowpea

Vigna sinensis

OFTEN the first thing aspiring growers want to know about cowpeas is whether they are peas or beans. While distinctly different from both, botanically the cowpea is closer to a bean. In the South, where they are most often grown, cowpeas are simply called peas, except for the dried seeds of the black-eyed variety which are called beans.

But whatever they are called, they are old. This annual legume dates back to prehistoric times. Originally from India, cowpeas traveled to Arabia and Asia Minor, and to Africa, where they still grow wild today. The crop is also known to have reached China before recorded history. Whether or not the early Greeks and Romans were familiar with cowpeas is uncertain. But by the fourteenth century, Italians were cultivating them, and they have continued to flourish in the Mediterranean region's ideal climate.

Most of the common plants introduced to the United States came here from Europe. Cowpeas did not. Carried on slave trading ships between Africa and the Americas, cowpeas were used on board to feed the slaves and were planted after landing in Jamaica. The warm climate was ideal and cowpeas spread throughout the West Indies, making their way from there to Florida around 1700. Cowpeas were first cultivated in North Carolina in 1714 and in Virginia in 1775.

When first grown in the United States, cowpeas were called "pease" or "callicance." An early custom of planting them between rows of field corn led to the name "corn-field pease." The name cowpeas first appeared in print in 1798. This name is the most common today, but is only one of many. Different varieties of cowpeas are called pea beans for their small size, black-eyed peas for their dark marking, crowders for those with seeds tightly crowded in the pod, field peas, and Southern peas.

There are several types of cultivated cowpeas. The tall climbing varieties (of which the asparagus bean is one) are cultivated most often in India and the Far East. They produce long pods, generally up to 16 inches. The tender pods of these varieties are usually picked young and eaten, beginning six to eight weeks after planting and continuing for about two months.

The cowpeas grown most commonly in America and Africa are a short, erect or weakly trailing plant, somewhat resembling bush beans. Pods develop on the top of the bush and grow between 3 and 12 inches long, containing as many as 20 small, tasty peas. The seeds are usually allowed to mature, and are used in their dried form. In Africa, and especially in regions of west Africa with higher rainfall, the mature seeds are the chief pulse crop and the seeds are often ground into meal. They are sometimes used as a substitute for coffee beans. Africans also eat the leaves of the plant which are cooked like spinach, and the immature pods—including underdeveloped seeds.

In the United States, cowpeas have typically been grown for purposes other than human consumption. As a legume, they add nitrogen to the soil, making them ideal for soil restoration. For this reason, they are often grown in areas where cotton and tobacco have depleted the soil. As a good green manure crop, cowpeas are plowed under to enrich the soil even further. These field peas are also cultivated as a hay crop or as forage for livestock.

But Southern peas are as nutritious for humans as for animals or soil. They have a high protein content, a unique hearty flavor, and can be eaten in diverse ways: as string or snap beans, shelled green beans, or dried beans.

Cultivated in tropical and subtropical regions all over the world, in the United States cowpeas are grown mainly in the southern states. But southern sections of Ohio and Illinois, New Jersey, and parts of Michigan are also suitable. Many areas in California are popular for raising cowpeas as well. They can be grown anywhere corn will grow, with enough warm days.

Cowpeas are very tender and easily injured by frost. "Planted too early they may catch a nip of late frost which they won't tolerate," observes Nancy Pierson Farris in South Carolina where she has grown cowpeas for years. "Summer plantings do well," she adds, "provided we have a long fall so they can produce before frost kills them."

Plant cowpeas the same time you plant your snap beans—after all danger of frost has passed. Make furrows leaving 3 to 3½ feet between them and plant the seeds 2 to 3 inches apart, and an inch deep.

(To grow cowpeas as a green manure crop, dig the field, broadcast the seeds, and rake to cover. You will need a quarter-pound of seed per hundred square feet.)

Cowpeas will grow in practically any type of well-drained soil. In fact, the soil should not be too rich or it will produce huge leafy plants and few beans. The plants are able to withstand drought very well, more easily than other legumes, and need little care. But watch out for brown stinkbugs which may attack the cowpeas, particularly when planted as a late crop. The best way to control stinkbugs is to keep the weeds down, but if they attack anyway, you can pick them off the plants by hand in the morning, or apply sabadilla dust (which is available from some suppliers of organic pest controls).

The different varieties of cowpeas among which you can choose for planting in the home garden vary in color and have sunbonnet-shaped blossoms of white, yellow, or lavender. They also vary in the number of days to maturity. The popular, sweet-tasting BLACK-EYED PEA takes 75 days. Newer early varieties of this favorite are ready in 50 or even 40 days. The PINK-EYED CROWDER is well suited to a short season, requiring only 50 days to mature, while other varieties take as long as 90 days. Standard varieties include the BROWN CROWDER, which produces long, round pods full of peas that turn brown when they're cooked or dried; the WHITE CROWDER, which produces white peas; the prolific, easy-shelling MISSISSIPPI SILVER; the wilt-resistant RAMSHORN BLACK-EYE; and the popular BLUE GOOSE. The LADY or WHITE LADY type is tiny, very prolific, and good for drying. BIG BOY and WHITE PURPLE-HULL CROWDER are two varieties that produce larger than average peas and are noted for their fresh flavor.

The best time to harvest cowpeas depends on your intended use. For string or snap beans, the pods should be harvested when immature, before the beans are fully developed in the pods. To use for shelled green beans, it is best to pick cowpeas when the color begins to fade and the seeds are nearly full grown, but before they've started to dry. Allow the seeds to mature and dry on the vine before picking them to use as dried beans.

To shell your cowpeas, break off the stem end, pull the string all the way down the length of the pod, pry the pod open, and remove the peas. You can cook green shelled cowpeas and immature pods together by breaking the young pods into convenient lengths and just adding them to the shelled peas.

The different ways cowpeas are used will determine not only when they are picked, but how they should be stored. Picked late they can be dried and kept indefinitely. Harvested earlier they can be canned in a pressure canner, or frozen like green beans. Blanch them for 3 minutes, cool, and pack to freeze.

A favorite Southern way to prepare cowpeas is to simmer them with a ham hock or some salt pork. When they start to lose their green color and are very tender, they're ready to eat. Cowpeas add body to soups of all kinds, and are a treat when cooked with onion or garlic until tender and then pureed with a bit of milk and topped with cheese.

Seeds for cowpeas are available from W. Atlee Burpee Co.; D.V. Burrell Seed Growers Co.; Comstock, Ferre, and Co.; DeGiorgi Co.; Gurney Seed and Nursery Co.; Jackson and

Perkins Co., Earl May Seed and Nursery Co.; Mellinger's, Inc.; L.L. Olds Seed Co.; Geo. W. Park Seed Co.; R.H. Shumway Seedsman; and Otis S. Twilley Seed Co. H.G. Hastings Co. and the Vermont Bean Seed Co. both offer especially good selections of Southern peas.

Hoppin' John

Here's a classic Southern dish that is traditionally eaten on New Year's Day for good luck.

1 pound dried black-eyed peas, washed
¼ pound bacon or salt pork cut in cubes, or a ham bone
1 onion, peeled and chopped
1½ quarts water
2-3 teaspoons salt
3 cups hot, cooked rice
2 tablespoons butter or bacon drippings
⅛ teaspoon crushed hot peppers (or more if you dare!)
⅛ teaspoon freshly ground black pepper

Put the peas, bacon, onion, and water in a kettle and simmer over low heat till the peas are tender, about 2 to 2½ hours. Drain most of the liquid from the peas, take out the bacon or ham bone, and stir in the rest of the ingredients. Serve hot, along with meat and cornbread. This recipe serves 6.

The Cresses: Winter Cress, Garden Cress, and Watercress

Barbarea spp., *Lepidium sativum*, and *Nasturtium officinale*

A Greek proverb promises that eating cress will improve your wits. For most of us, that's an appealing prospect, and so are the pungent leaves of the various cresses belonging to the Crucifer family. These plants are usually treated separately as watercress and as the botanically different kinds of land cresses, which themselves belong to two different genuses.

WINTER CRESS is the general name given to the various *Barbarea* species which were named after St. Barbara in the ancient belief that eating these greens on that saint's special day (December 4) would bring good luck. So-called COMMON WINTER CRESS (*B. vulgaris*), often found growing wild, is also known as spring cress and yellow rocket. It's also sometimes called upland or land cress, though these names are more accurately used to describe *B. verna*. A biennial that is more often an annual or winter perennial, the hardy, low-growing common winter cress has dark green leaves that form a rosette pattern and look rather like those of watercress, though they're somewhat thinner and less succulent. Having from one to four pairs of lobes per leaf and a large, rounded end lobe, common winter cress differs in this respect from the more refined UPLAND or LAND CRESS, which can have from five to ten pairs of side lobes. Both plants, however, grow to 8 to 30 inches and send forth small clusters of edible bright yellow flowers followed by pointed, 1½- to 2½-inch-long seed pods that supply good bird seed. Winter cress greens taste like watercress when raw, but more like mustard greens when cooked.

These plants are widely distributed along streams and in roadside swamps in North America, especially in the north central and northwestern states and in eastern Canada. If your winter climate happens to be mild, you can sow winter cress from July to September for winter and spring use. It will mature in eight weeks and you can cover it with mulch over the winter to improve its eating quality. In colder climes, plant in well-manured or composted soil as early in spring as you can work the earth. Cover the small seeds only lightly—or just put a board over them for four or five days until they germinate. Later, thin them to 4 to 6 inches apart.

You'll find winter cress does well in north-facing borders and in succession plantings in damp, semi-shaded spots, interplanted between tall or widely spaced crops. The plants will survive a few hard frosts, and if you want to keep enjoying their lively flavor, you can bring them indoors in pots in October. Make sure, though, that you keep them in an area with a temperature under 68°F.

Harvest the leaves as needed, eating them raw or as a steamed green while they're young, and parboiled for 2 to 3 minutes, then drained and cooked again if they're older. You also can blanch extra winter cress and freeze it with some cooking liquid for out-of-season enjoyment. A variant of common winter cress is EARLY WINTER CRESS (*B. praecox*), a native of Europe also called Belle Isle cress and scurvy grass. It is used just like other winter cresses.

Winter cress makes an intriguing addition to quiche—cook about a cup of cress, drain, and spread on the bottom of the pie shell along with the grated cheese you usually use. A lovely salad can be concocted by combining winter cress with other in-season greens. In spring, try daylily shoots and violet leaves; later on add dandelion, chicory or corn salad.

GARDEN CRESS (*Lepidium sativum*) is also known as curled cress and sometimes as pepper grass, though some botanists say that pepper grass is a general name used for a number of *Lepidium* species—many of them wild plants native to the United States. (Just to make matters more confusing, you might like to know that famed horticulturist Liberty Hyde Bailey classified pepper grass as *L. virginicum*!) An easy-to-grow annual, garden cress is thought to have originated in western Asia, and was a popular garden vegetable in ancient Greece. Today it's raised all over Europe, including the British Isles, and is often germinated on a plate of moist cotton, since it grows phenomenally fast and is quite tasty when eaten as a tender seedling. Sometimes crisped like parsley, garden cress leaves are small and green, varying greatly from species to species and even from habitat to habitat, which might explain the confusion in nomenclature. Upright and single stemmed or sparsely branched, the plant grows anywhere from 8 to 24 inches tall, producing white or reddish flowers. These are followed by tiny papery pouches containing hot seeds that add a special tang to soups and salad dressings.

A cool weather plant, garden cress flourishes in most temperate regions all over the world in clearings, fields, and along roadsides where the soil is somewhat dry. Because this cress grows so fast, it peaks quickly and goes to seed faster than winter cress. Moreover, with the coming of hot weather, its flavor gets quite peppery. You can grow garden cress indoors in bonsai trays, pots, or flats on an east- or south-facing windowsill where the temperature is under 68°F. Or raise this cress in cool rich soil in the corner of a cold frame or in some light compost at the edge of a garden bed. Sow your seed thickly and cover lightly or not at all, providing water and shade when needed. Keeping the rows 9 inches apart, thin the plants to 3 inches and try succession plantings every two or three weeks. To grow this tangy green in a greenhouse, sow three or four seeds to the inch and thin the seedlings to stand 1 inch apart. Plant under the benches since cool rich moist soil is essential and sun is not.

To savor garden cress at its best, harvest the leaves and pods while the flowers are still on the stalks. If you're not interested in succession planting, you can cut leaves for salad when the plants are 6 inches or so high, and you will get

• I have a concrete bird bath under a juniper bush, and I often plant some of the watercress with roots along the lower stems in the gravel at the bottom of this bird bath. It often thrives there for a month or more.

—*Catharine O. Foster*
Bennington, Vermont

another cutting before the leaves turn hot if you don't damage the plants' crowns.

This cress is a very fine source of vitamin A, offering an excellent 9300 international units per 100 grams—almost twice as much as watercress. Garden cress also has good amounts of vitamin C, iron, and calcium. Try the leaves in salads, in soups, and in sandwiches, along with sliced hard-boiled eggs, or enjoy the seedlings that way as the English do. Try some in scrambled eggs or an omelet. Use the whole seed pods fresh or dried as a seasoning and save the dried seed stalks as a handsome soft beige addition to dried fall arrangements.

WATERCRESS also belongs to the Crucifer family, though it bears the rather misleading botanical name of *Nasturtium officinale*. This aquatic plant is a perennial native to Europe and western Asia. Gathered wild from streams for most of its history, today watercress may be searched out throughout the United States in streams, in spring-fed ponds, and near natural streams. In olden times, the Romans reportedly ate watercress with vinegar to help cure mental problems. It also has been used as a cure for falling hair and scurvy, and was viewed by the Persians as an outstanding food for children. Sometime around 1800, market gardeners in what is now Erfurt, Germany, began to do a brisk business growing watercress in long wide beds near natural springs. An enterprising French hospital administrator carried the idea back to France, and watercress was on its way as a commercial vegetable—one that now commands premium prices from city folks addicted to its crunchy greenery and to a zingy flavor that leaves no afterburn.

Prostrate or trailing, watercress has many-lobed leaves and thick, fleshy stems with lots of root hairs. At its stem tips grow groups of small and inconspicuous white flowers, the appearance of which announces a stronger, sharper flavor in the cress. This plant matures in 50 days and is in season from late spring to the end of November in temperate places, and year-round in mild climates. In Florida, home gardeners can grow watercress from October to May, when it will reseed and come up again in fall.

If your property is graced by an unpolluted stream, you

• To prepare "cressy sallet," I cut off winter cress right at the top of the ground, close to the first leaves. I wash it many times to remove soil, other seeds, grass and dirt. Then I boil till tender and remove from water into bacon grease, salt it and cook a few minutes, then take up and serve with crisp bacon and cornbread. Mmmm, mmmm good!

—*Mrs. Nolan Mitchell*
Florence, Alabama

can start some plants from seeds or from healthy supermarket sprigs rooted in water and go on to establish a bed or two in handy elbows of the stream. (Do remember that watercress likes shady, woodsy areas.) If you don't happen to have a stream passing by, you can make your own landlubber's bed for this gourmet treat. First, dig a 6-inch-deep basin with a level bottom, keeping the area around the outer edges flat. Fit in a plastic sheet, then fill the basin to a depth of 2 inches with a blend of peat moss, phosphate rock, and granite dust. Soak this mixture thoroughly, then scatter seed at one end of the bed and cover with a light sprinkle of granite dust. Keep moist while the seeds germinate—this should take about five days—then raise the water level as the plants grow. When they're about 2 inches high, transplant them throughout the bed at 6-inch intervals. Fertilize with fish emulsion a few times during the season. You also can grow watercress indoors in plastic flats on a windowsill or in a greenhouse (or in a hotbed or cold frame) if you plant in sand and peat moss, fertilize occasionally, and keep under an inch of water. Those who try this method inside suggest flushing the setup once a week by letting water trickle through slowly in order to avoid the perils of stagnant water.

If you wish, you can sprout watercress seeds as you would alfalfa, or broadcast them in a container for a European-style harvest of sprouts. If you prefer grown-up watercress, be sure to harvest your fast-growing plants every 15 days or so to keep the leaves from getting tough and scorching hot. Your crop will please you with its good food value, for 100 grams of raw stems and leaves contain 4900 international units of vitamin A (almost twice as much as broccoli), more calcium than half a glass of milk, and as much vitamin C as an orange.

Try watercress in salads or chopped and blended with butter to make a tangy sandwich spread. It also tastes scrumptious in stuffings, omelets, mashed potatoes, dumplings, cream cheese, or cottage cheese. For a garden-fresh watercress soup, clean and chop coarsely 1 large bunch watercress, 1 bunch parsley, 2 green onions with tops, and 2 or 3 new potatoes (with skins on, if organically grown). Drop the vegetables into a quart of boiling water, cook until the potatoes are done, then cool slightly and whiz for a few seconds in the blender. Return to the pan and add either a cup of chicken stock or a cup of light cream and a little salt and pepper to taste.

You can order the various cresses from a host of seedsmen. These include W. Atlee Burpee Co. (Curly, Salad, Watercress); Casa Yerba (Garden and Watercress); Comstock, Ferre and Co. (Pepper Grass, Upland, Watercress); William Dam Seeds (Garden); DeGiorgi Co. (Curled, Upland, Watercress); Gurney Seed and Nursery Co. (Curled); Joseph Harris Co., Inc. (Garden, Watercress); Charles C. Hart Seed Co. (Curled, Watercress); H.G. Hastings Co. (Upland, Watercress); J.L. Hudson, Seedsman (Garden, Upland, Watercress); Le Jardin du Gourmet (Garden, Watercress); Johnny's Selected Seeds (Watercress); Meadowbrook Herb Garden (Garden, Winter); Nichols Garden Nursery (Upland, Pepper Grass, Watercress); L.L. Olds Seed Co. (Curled, Watercress); Geo. W. Park Seed Co. (Garden); Redwood City Seed Co. (Garden, Upland); Seedway, Inc. (Garden); R.H. Shumway Seedsman (Garden, Upland, Watercress); and Stokes Seeds, Inc. (Fine Curled, Watercress).

• For a luscious one-dish meal, cook 3 or 4 potatoes and your favorite kind of sausage (cut in bite-sized pieces) in water to cover until tender, adding salt and pepper to taste. Drain off the water, and add a cup of milk and a bunch of watercress, chopped. Mix together thoroughly. Cook another 6 minutes, top with a bit of butter, and dig in.

Chinese Cucumber

Cucumis melo

TEN years ago, John Meeker walked into a market and saw a colorful seed packet imported from Japan. In Japanese the packet announced these seeds to be "Yard-Long Cucumbers." Like everyone else who is a sucker for the-bigger-the-better, he was immediately taken in by this promise and snapped the seeds right up. After planting some he found that these extra-long cucumbers, though not nearly a full 36 inches, were indeed very long, becoming regularly 18 inches in length. Growing them, he was pleasantly surprised to discover some other features: they are delightfully crisp, very easy to eat, not at all bitter even with the skin left on, and they do not cause stomach turmoil as some cucumbers do.

Little is known about the origins of the long cucumber, most commonly known as the Chinese cucumber. Also called Japanese, Oriental, Sanjaku (3-foot), Armenian, and yard-long cucumbers, they originated in southern Asia and have been cultivated in India for centuries. From there they spread to the West where the ancient Greeks and Romans are known to have enjoyed them.

Basically, Chinese cucumbers grow like other cucumbers but their fruit is longer and thinner, typically between 1½ and 3 feet long and a mere 2 or 3 inches wide. These varieties have a thin, bright green skin, few spines and seeds, and flesh that is

155

• Kids love to watch Japanese cucumbers grow. Let each kid in the family give theirs a name and they'll inspect it daily to marvel at its growth. They may not eat it, however, when it reaches time to pick. Kids have a sense of loyalty to their friends, even vegetable friends, that adults could emulate.

—*John Meeker*
Gilroy, California

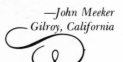

firm, white and very crisp. The Chinese cucumber is highly resistant to disease. Even gardeners who have had trouble with cucumber wilt with other varieties have had success with the Chinese cucumber.

There is one characteristic of the Chinese cucumber that is unique—a tendency to twist and curl. "If in the process of stretching out," observes Lewis Theiss of Pennsylvania, "it noses up against any obstacle like a small stone, it turns away and begins to curl. I have seen them assume the shape of a horseshoe. But that is an extreme example. There is one way to get around this: hang the cucumbers on a fence. Gravity pulls the fruit downward and keeps them straight."

But since these varieties have a strong tendency to climb, trellising them is simple. Not only will you save ground space, but growing them vertically will help attract pollinating insects and in turn benefit your entire garden.

Chinese cucumbers should be planted in the spring, after all danger of frost has passed and the soil is thoroughly warmed. A good garden loam, rich in humus, and with a pH of below 6.5 is ideal for Chinese cucumbers. If your soil is too acid, add some ground limestone or phosphorus to achieve the proper balance.

These cucumbers need plenty of sun, particularly morning sun. Seeds should be planted about an inch deep directly in the soil since cucumber roots are sensitive to injury during transplanting. Sow seeds about an inch apart in hills or rows and gently cover with damp soil. Thin seedlings to stand about 1 foot apart after the plants are established.

If you do not have an existing fence to support Chinese cucumbers, try growing them on chicken wire, as Betty Brinhart of Massachusetts has done. She puts 8-foot-high two-by-fours in the ground, 6 feet apart, and stretches the chicken wire between them. With rows running east to west, the large leaves grow toward the sun and the cucumbers have plenty of room against the chicken wire.

Be sure to give Chinese cucumbers enough water. The fruit is 96 percent water, which can give you an idea of how much is necessary. Growth will be hampered if the plants are not kept moist: they will bolt, one end becoming disproportionately large and the other end shriveled and immature. But with sufficient moisture, Chinese cucumbers will bear until frost. Mulching with old hay, oak leaf mold, or other organic materials helps maintain the moisture and temperature and permits even growth. It also keeps down weeds.

During their growing season, Chinese cucumbers can be harvested when still small for pickling or relishes. The Japanese pickle them in bran or *miso* culture, but they're good for pickling by any method.

For most use, however, wait to harvest until they have grown out. Lucille Eisman of Ohio recalls picking her first Chinese cucumber when it was about a foot long. It was tasteless. A hurried phone call to the friend who had given her the seeds taught her patience.

"Goodness, only a foot long!" her friend exclaimed. "Wait until they're 18 inches, at least, and the skin should have a slight yellowish tinge."

Within a week Ms. Eisman's cucumbers were 20 inches long, yellowish green, ripe, and very sweet.

Unlike many vegetables that lose their sweet taste when large, Chinese cucumbers get sweeter. And even when full grown, they still have very few seeds. They are best picked when the length has filled out but the spines are still somewhat prominent. Pick them regularly to strengthen the plant and improve the yield.

Chinese cucumbers do not store very well. They become limp and lose their crispness. So harvest them as close as possible to the time they will be eaten.

These large cucumbers can be used and prepared in the same way as their smaller relatives. They are delicious raw, skin and all, and produce no gas pains. In Japan it is more common to peel them before eating. Chinese cucumbers make an excellent addition to salads or sandwiches, and their shape makes them very suitable for slicing. Also, try grating Chinese cucumbers into soups, frying them, broiling them, or serving them with a white sauce. They're delicious stir-fried and seasoned with soy sauce and chopped fresh chives. A cold soup

• Many Japanese have never seen a fully grown (Chinese) cucumber. Just for fun we sometimes carry one of these huge vegetables with us on the streets of Kyoto City when we visit friends there. It never fails to astonish and amuse (in that order).

—*Larry Korn*
San Francisco, California

made with Chinese cucumbers and yogurt mixed in a blender is a refreshing treat on a hot day.

As a food, Chinese cucumbers are valuable for their sugar and vitamin content, and their digestibility. But the beauty of the plant's female blossom makes it a splendid decoration floating in a bowl of water.

Seeds for these long cucumbers are available from Burgess Seed and Plant Co.; W. Atlee Burpee Co.; Glecklers Seedmen; Grace's Gardens; Joseph Harris Co.; J.L. Hudson, Seedsman; Johnny's Selected Seeds; Kitazawa Seed Co.; Lakeland Nurseries; Nichols Garden Nursery; Redwood City Seed Co.; and Stokes Seeds, Inc.

Oriental Salad Bowl

This dish is nice as a complement to a Chinese or Japanese main dish, or just as a change from your usual tossed salad. In a large salad bowl, place 3 cups of mixed greens—shredded Chinese cabbage, chopped Shungiku—whatever happens to be in season. Add about a cup of oriental cucumber peeled and cut in matchstick strips, some coarsely grated or sliced winter radish, shredded burdock root or carrot, and some tiny cooked and peeled shrimp. Make a dressing of 2 tablespoons sesame or peanut oil, 2 tablespoons soy sauce, 2 tablespoons rice wine vinegar, a tablespoon of sherry, if desired, and salt and pepper to taste. Toss all together and serve.

Dandelion

Taraxacum officinale

DESPITE a rather ferocious name meaning "lion's tooth" (from the French *dent de lion*), this plant with the jagged leaves is an untemperamental fellow associated in folklore with lovers eager to know their romantic fate. According to tradition, if you can blow all the seeds from a downy dandelion head you are loved with passion. If a few seeds remain, your lover has some second thoughts about you and may be unfaithful, and if many seeds stay put, you're loved rather tepidly or not at all. If this experiment doesn't work out to your liking, you can console yourself with another bit of folk wisdom that says the airy puffs you blow off are carrying your thoughts to your lover.

Believed native to Eurasia, the dandelion is a perennial that has been eaten in Europe and Asia for centuries. It now grows in the temperate zones of all the continents, thriving in any climate that will support spinach or kale. More often than not, alas, the dandelion's long taproot and somewhat bitter leaves have earned it the reputation of being an unusually persistent weed—particularly in the lawn-loving United States. During wartime, however, the humble dandelion has always emerged as a much-valued vegetable, and since the end of the last century it has begun to find its way to the marketplace. In America it is often savored by people who relish Italian,

161

Pennsylvania Dutch, or other Old World styles of home-cooking.

Shiny, smooth, and rich green in color, dandelion leaves develop in the form of a spreading rosette atop a very short stem. This is attached to a long root that may reach a depth of 3 feet. Because the dandelion is an aggregate species, plants in different places are considerably varied in the size and shape of their leaves and in the form of the bracts at the base of the many-petalled yellow flower heads. Some varieties and strains look like endive, which is also a member of the daisy family.

Several cultivated varieties of dandelion now available tend to have larger leaves than wild dandelion. The large-leaved French variety is preferable to common dandelion, but better yet is an improved broad-leaved variety that doesn't go to seed as quickly as the French. Called THICK-LEAVED DANDE-LION, this plant has broad leaves more deeply lobed along the axis of the leaf stem than those of wild dandelion.

In the United States, dandelion will grow just about everywhere except the hottest parts of the South. Sow the seed in deep-spaded soil as early in spring as the ground can be worked. Or plant in midsummer to start plants that will yield masses of tasty greens the next winter or early spring. (These fall sowings should be covered with a straw mulch over the winter in the North, but they'll need no protection in the South.) Dandelions take care of themselves in most any kind of soil, including the sandy sort, but they'll grow faster and produce superior leaves if you pay them the compliment of deep rich loam kept well cultivated.

When you choose the site for your dandelion patch, keep in mind that this vegetable gives off ethylene gas, which tends to limit the height and growth of neighboring plants and causes nearby flowers and fruits to mature early. Sow your seed thinly in furrows ¼ to ½ inch deep in rows from 18 to 24 inches apart, and keep the soil from drying out during the germination period. Thin early because of that tenacious taproot, and leave about 10 inches between plants, though you can crowd them as close as 4 inches if you plan to harvest the plants whole before they reach full growth. This can be a good idea, for though the seed packets say dandelions take 95 days to maturity, if you let them go that long you'll probably find them pretty tough customers. Besides, who wants to wait? The first tender inner crown of new leaves hiding under winter mulch or pushing through the frosty ground in April makes a welcome salad green, and the leaves of new plants are at their flavor peak about six weeks after planting. At that time you might want to harvest them roots and all and start a new batch.

The alternative to such an early and absolute harvest is to let some or all of your plants go the full 95 days—*after* you tie up the outer leaves of each to blanch the inner "heart" a golden yellow. If you like endive, this could be the way to go, for the blanched leaves will be very tender and less bitter than older green leaves. On the other hand, you will lose much of the astonishing vitamin A value and respectable vitamin C content that make dandelion so worthwhile.

If you haven't completely used up your first or second planting at the end of the growing season, you may want to lift the plants roots and all for forcing. This can be done in containers in the garage or cellar, or in a greenhouse or hotbed. You can even do it outdoors, using a hole filled with sand as your container. Just wash off the roots and cut away all the dead and outer green leaves, leaving only the youngest leaves at the crown. Pack the roots in damp sand, then cover the crowns with the same damp sand to a depth of about 8 inches. Finally, put an upside-down cardboard box or other lightproof cover over your pot, box, or hole. If you want fast forcing of growth, put the container near a source of heat (such as a water heater). Be warned, though, that the heat may also make the faster-developing leaves more bitter. Within 10 days or so—if the forcing area is warm—the dandelions will pierce the top of the sand with growing leaves. At that point you can harvest the leaves as they appear for two to four weeks, then empty out the sand and enjoy the rest of the white and tender-crunchy forced leaves in winter salads.

If you want the more vitamin-rich version of dandelion for winter eating, you can grow plants in a greenhouse. Sow the seeds in rows about two seeds per inch and thin the seedlings to stand about 6 inches apart. You can also blanch plants under glass by tying them or covering them with a tarpaper collar to exclude light.

As all these options suggest, you can enjoy this hardy green through much of the year. A closer look at its nutritional profile should persuade you to do just that, for the much-maligned dandelion is one of the vegetable kingdom's most potent sources of vitamin A. Just a single cup of the raw greens packs in 14,000 international units—that's over five times as much as you'll find in broccoli, almost twice as much as in collards, over 1/3 more than in raw spinach or cooked sweet potato, and 1/5 more than in the same amount of raw carrots! The mighty dandelion is also chock full of iron, offering almost as much as raw beet greens, which excel all other vegetables except beans and lentils. Raw dandelion also contains handsome amounts of calcium and vitamin C and is much higher than most green leafy vegetables in the B vitamins thiamin and riboflavin.

You can enjoy younger green or forced leaves as part of a spring salad, together with rocket from your garden, lambsquarters, pigweed tops, mustard greens and other hearty wild foods. Dress this zesty treat with a simple vinaigrette sauce made of oil, salt, pepper, and wine vinegar. Or try a memorable blend of two parts olive oil to one part lemon juice.

Young or older leaves can also be boiled in a little water along with young sorrel leaves till tender, then served with butter and salt and pepper either alone or with poached eggs. For a superb-tasting vitamin A high, you might try the same approach with a mixture of dandelions, beet tops, and collards. Or you can chop and steam dandelions, then add them to omelets, fritters, and quiches. If wilted lettuce is your idea of good eating, chop up 3 or 4 cups of dandelion and mix it in a salad bowl with 4 or 5 crumbled slices of well-cooked bacon and 2 finely chopped hard-boiled eggs. Next, thoroughly mix 2 heaping tablespoons of mayonnaise with ½ cup vinegar and heat to a boil in the pan still holding the bacon grease. Pour this dressing over the dandelion mixture slowly and toss the salad. Serve hot to three or four dandelion lovers.

To enjoy more mature dandelion greens, boil them for

• Being a dandelion greens lover and regretting that they last only a couple of weeks in early spring in the wild state, I bought seeds from a nursery some three or four years ago and planted them in the garden. They are still there and in the same rows where they were planted, as I don't plow or till up the garden any more.

Each spring I loosen the soil and plant around them, then harvest the greens all spring, summer and fall. Besides, I always freeze the winter's supply out of probably 25 plants in all. Last summer they self-seeded a few plants, which I'll take up and put in the row this spring.

All summer, I dump the

about half an hour, then drain them very thoroughly and chop fine. Add the leaves to a saucepan in which you've mixed together 1 teaspoon of unbleached or whole wheat flour, 1 tablespoon of butter, 1 tablespoon of vegetable or chicken stock, and a pinch each of pepper and salt. Heat for 10 minutes or so, then stir in a gob of sour cream, yogurt, or some half-and-half and serve. You can always braise tender or blanched dandelion leaves in butter, then add a touch of tarragon and lemon. For variety, try a chopped dandelion sandwich on whole wheat bread with plenty of butter and mayonnaise. Or make a sandwich spread or vegetable dip by whizzing dandelion leaves with some cottage cheese, chopped nuts, and mayonnaise in the blender.

Dandelion flower buds make a tasty addition to an omelet. Or you can batter-fry them or make them into fritters. As every connoisseur of home brews can tell you, the flower heads of this useful plant can be used to flavor and color a delectable wine. And you can convert its scrubbed and dried roots into a coffee substitute by roasting them in a 200° to 250°F. oven until they're dark brown. This will take from three to five hours. When you

---

DANDELION WITH HOT BACON DRESSING

Here's a rich, traditional Pennsylvania Dutch way to serve dandelion. Start by frying, draining, and crumbling 4 slices of bacon. Set them aside, along with 3 or 4 cups of washed, chopped, and dried dandelion greens. Warm ¼ cup of butter and ½ cup of cream, milk, or water in a skillet, over low heat. Meanwhile, mix 1 teaspoon of salt, a dash of pepper, dash of paprika, 1½ teaspoons of honey, and ¼ cup of cider vinegar into 2 beaten eggs. Blend this mixture into the warm butter and cream in the skillet and cook over medium heat, stirring, until thick. Stir in the dandelion and bacon, cook for just a minute to wilt the greens, and eat promptly. This recipe will make 3 or 4 servings.

---

dry garbage and wastes right on top of the leaf and straw mulches, whereas the smelly type of garbage is hidden underneath. My earthworm population is sufficient to take care of such practices—and a good thing too, as I don't have the physical strength to do much composting.

The dandelions will be much earlier this spring for having had this coverlet of leaves and straw through the zero weather. I uncover them by degrees and will probably find a mess of dandelions ready under the mulch in early April, possibly in March.

—*Gertrude Springer*
Delton, Michigan

grind and brew these roots, you'll have a noncaffeinated drink with the properties of a tonic and stimulant. By the way, if you dig them up when young, dandelion taproots can also be boiled and eaten as a vegetable. Season them like parsnips.

Seeds for improved dandelion varieties can be bought from W. Atlee Burpee Co.; William Dam Seeds; Charles C. Hart Seed Co.; H.G. Hastings Co.; J.L. Hudson, Seedsman; Le

SOME DANDELION ANTICS

It's a tough plant, the dandelion. But there are some odd facts about its habits that are worth knowing. For one thing, the dandelion is a short-day plant. That is, it won't bloom whenever there are more than 12 hours of light (or less than 12 hours of darkness). That applies to the majority of the plants in a given location, though, so a few plants can be seen blooming in the long summer days of July.

The dandelion flower remains open for only one day, on the average, then closes again. After blooming, the closed flower and the stalk it is mounted on flatten to the ground. After another couple of days, the scape (stalk) straightens itself like a rising periscope and the leaves surrounding the closed bloom reopen to reveal the familiar round white ball. This contains feathery seeds being readied for distribution.

When the dandelion's large storage root is injured or broken near its upper portion, it grows a callous tissue to plug the wound. After that, one or more buds are formed on the tissue and either new leaves or an entire new plant develop. For instance, when a grazing animal plucks a plant by tugging at the leaves, the usual result is that the upper third of the root breaks off. The part remaining in the ground grows a callous tissue to close the wound, and eventually two to five new plants can arise from this tissue! That's why you often see dandelions growing in rings. In such cases, there probably was a single plant in the center ring, and on losing the rosette of leaves above ground, the remaining root produced a series of new plants.

Jardin du Gourmet; Meadowbrook Herb Garden, Mellinger's, Inc.; Nichols Garden Nursery; Redwood City Seed Co.; R.H. Shumway Seedsman; Stokes Seeds, Inc.; Thompson and Morgan, Inc. and Well-Sweep Herb Farm.

Daylily

Hemerocallis, spp.

ORIGINALLY brought to this country from the Orient, the daylily escaped from cultivation long ago to grow along roadsides, hedgerows, and near old homesites from Canada south throughout most of the United States. Their bright orange, yellow, pink, melon, and red blossoms of the many recent cultivars make them favorite perennials among country gardeners, but as a spring through fall source of elegant, oriental vegetables, they deserve a spot in the vegetable patch as well. The young sprouts that appear in the spring, the summer buds and blossoms, the leaves and even the rhizomes are considered delicacies by gourmets and wild food gatherers.

There are about a dozen species and several thousand named varieties of daylily (*Hemerocallis*), all of which differ from the true lilies (*Lillium*) in that they do not produce bulbs. The name *Hemerocallis* is derived from the Greek words *hemera,* meaning day, and *kallos,* meaning beauty. Each blossom survives only one day, but by planting early-, midseason-, and late-blooming varieties, flowers can be had from spring till fall. Each flower has six slightly overlapping petals that curve backwards as the flower opens to the sun.

Fine, rugged perennials, daylilies are easy to grow in almost any kind of soil, in sun or half-shade. Hybrid varieties and the so-called "lemon lilies" grow into clumps which must be

169

lifted and divided every few years, in spring or fall, to keep them blooming well. The tawny daylily, *Hemerocallis fulva* (the common orange species found almost everywhere), spreads by underground stems or rhizomes to form broad patches in the garden or in the wild. Daylilies are hardy in most climates, and establish themselves quickly in the garden. They require almost no care except an occasional top-dressing of compost or leaf mold to insure heavy flowering.

The first harvest of daylilies takes place in early spring when the young leaves appear. Although the plants are virtually indestructible, it is best to cut only the outer leaves, avoiding damage to the flowering stalks. These leaves are especially tasty and tender when they are 3 to 5 inches long and still young. They can be simmered or stir-fried in oil or butter, and some folks compare the taste to creamed onions. Eaten in large amounts, the leaves are said to have a mild hallucinatory effect; they were used by the Chinese as a pain reliever. The flower buds and blossoms are often considered the most delectable part of the plant. They appear in midsummer and can be eaten at all stages of their growth. Their slightly flowery sweetness and mucilaginous texture add interest to soups and vegetable dishes. The tightly closed buds can be used in salads, boiled, pickled, or, like the leaves, stir-fried. Try them steamed with snow peas, or stir-fried with pork, onion, and soy sauce. Half-open, fully open, and even one-day-old daylily blossoms can be dipped in a light batter of flour and water and fried in a wok, tempura style. They are delicious in omelets and other egg dishes, and go well with chicken, too. The Chinese dry the young buds and open flowers and call them "golden needles." To dry your own, simply string them on a heavy thread and hang in a high, dry place for a week or two until thoroughly dry. Then store them in an airtight container. Or, place on a drying rack for several days, turning at least once a day. Before using, soak the buds or flowers in warm water for several minutes.

The plant's tiny tubers are also edible at practically any stage of growth. But in late fall or winter, when they have stored food over the summer, they are at their best. This is a

• My sister-in-law, who is from Germany, served me daylilies gathered from a roadway near Ithaca, New York. They were delicious. I've also eaten them from the landscaped area between the restaurant and the motel office at Howard Johnson's in San Jose, California. I am no Euell Gibbons but I am an inveterate daylily muncher. The other day when one of my small grandchildren reminded me of a pleasant family outing which had occurred, she said, "the day we learned to eat flowers." She reminded me that I have been passing on the habit to my children.

—*John Meeker*
Gilroy, California

good time to dig up and divide overgrown clumps that need rejuvenating. While you're doing that, select a number of firm, white tubers, scrub them, peel, and use in whatever way you like. Daylily tubers are very crisp, with a wonderfully nutty flavor, and can be prepared in many of the same ways you serve potatoes. They can be boiled and creamed, eaten raw as a snack or in a salad, or even added to all kinds of soups and stews. For a special treat, try them cooked until tender, mashed and mixed with bread crumbs, salt, pepper, chopped parsley, and a bit of milk, and made into patties to be browned in oil.

Daylilies can be obtained from almost any nursery that stocks perennials, and from many seed houses as well. Among the larger suppliers are W. Atlee Burpee Co.; Geo. W. Park Seed Co.; and Thompson and Morgan, Inc. Gilbert H. Wild and Son, in Sarcoxie, Mo. 64862, specialize in daylilies and have an enormous selection of hybrids and standard varieties.

Daylily Chop Suey

Crunchy daylily tubers are a welcome addition to chop suey. To make 3 to 4 servings of this tasty dish, stir-fry a chopped onion in 1 to 2 tablespoons of oil until tender. Add 2 boneless chicken breasts cut in bite-sized pieces, and stir-fry until the chicken is white and cooked through. Then add a cup each of chopped celery, green pepper, mushrooms, bean sprouts, and peeled, sliced daylily tubers, and stir-fry briefly. Add a cup of chicken broth and soy sauce to taste, turn down the heat, and simmer until the vegetables are tender, 5 to 10 minutes. Serve over hot rice.

Japanese Eggplant

Solanum melongena

THE Japanese call eggplant *nasu,* and the eggplant commonly grown in Japan looks and tastes different from the vegetable with which most of us are familiar. Looking like a miniature mutation of the glossy, global American and Italian varieties, the Japanese eggplant has a milder, more delicate taste. Even the most patriotic of gardeners will generally concede that the smaller *nasu* possesses a much more delicate and distinctive eggplant flavor than its bulbous American counterpart. The mature *nasu* is elongated, slender, and about 6 inches in length. Its shiny purple black skin is thinner than the skins of most other varieties and consequently less tough.

Japanese eggplants generally grow best in hot weather, although one Montana gardener found that, if covered with blankets when frost was expected, her Japanese eggplants kept on producing in cool autumn temperatures. The plants need full sunlight and plenty of water, but at the same time, it is equally important not to overwater them. Too much water tends to expand the size of the fruit while at the same time depleting its flavor. Proper drainage is another necessity for a successful crop. In Japan, where poor drainage is a constant problem, eggplants, like most other vegetables, are grown on hills with drainage trenches between the plots. These raised permanent beds serve a dual purpose. Besides providing the plants with adequate drainage, they also keep the gardener's feet from trampling on their sensitive roots. A California gardener says in past years he's had bad results with eggplants which he could only trace to his walking between rows too near the roots of the plants.

Because eggplants as a rule react poorly to cold weather and take over three months to produce, it's advisable to start your plants indoors. Plant seeds ¼ inch deep in small containers, such as peat pots or cut-down milk cartons, filled with a good, rich potting soil. Water well and place in a warm spot. One gardener's advice is to cover the containers with a plastic bag, then place them in a cupboard above the oven. If this is inconvenient or you don't have a cupboard located over the oven, a warm window with a southern exposure will suffice. Another way to ensure warmth for the seeds is to place an old heating pad under them. The seeds should germinate in between 7 and 10 days. When the second set of true leaves appears, you might transplant the young plants to peat pots to help prevent their suffering root shock on their journey to the garden site.

If started in March, the seedlings should be ready to set out to harden off in a cold frame by April in most locations. After the danger of frost has passed, place them near a sunny wall for a few days to harden off further.

The Japanese eggplant has trouble developing a sturdy root system, so early planting is a must. Remember this variety will not grow as large as its American relatives, so placing the plants between 15 and 18 inches apart is ample room.

For the first few weeks the overanxious gardener is likely to rue the day he was ever convinced to experiment with this foreign vegetable. *Nasus* are slow in getting started. Like tomatoes they just won't budge until the nights are fairly warm and the days good and hot. Those gardeners who are familiar with oriental philosophy know that patience is nearly always rewarded, and they'll find that once the soil becomes thoroughly warmed the plants will take off like a kite on a windy day.

Cool weather may be detrimental to the eggplant's productivity, but equally damaging is poor soil. Because the plant is a heavy feeder with a very high potassium content, you need to provide enriched soil well fed with decomposed organic waste. A side-dressing of fish fertilizer or compost in late summer will also give the plants a needed boost.

Like Chinese cucumbers and tomatoes, Japanese eggplants appreciate a light mulch. Not only will mulching serve, as usual, to conserve water and control weeds, but it will also guard against a disease to which eggplants are sometimes susceptible. This disease, which usually appears in midsummer, causes damage in the lower half of the plant when sunlight reflects off the surface of the ground. The leaves yellow, followed soon after by insect attacks. This happened to a gardener who used to live in Japan. After consulting the local expert (a grandmother next door) he was advised to mulch his ailing plants with rice hulls. Both the disease and the insects were gone in ten days. Other knowledgeable gardeners suggest placing a black plastic mulch or aluminum foil around the base of each plant to guard against slugs.

When the plants sprout their modest violet blossoms, it is the first indication that it won't be long before *nasus* are on your table. It is wise to pick off the first few flowers so the plant will conserve its energy and concentrate on more important matters, namely fruit production. Experience has shown that eggplants produce more and for a longer time if these early flowers are not allowed to bear fruit.

After 90 days the *nasus* should be shiny, purple-black and ripe for the picking. If you notice the glossiness dulling on the eggplant, pick it right away. The dullness means the eggplant is past its prime. It may be slightly bitter, but is nevertheless still perfectly edible. The fruit is ideal when around 6 inches long. About 10 of these plants should furnish plenty of vegetables for a family of four.

Try to use your Japanese eggplants soon after you pick them. They do not store well, or lend themselves to freezing. Even after a few days of refrigeration, their quality and appearance begin to deteriorate. If you find that you do need to find a way to store them, you'll have the best luck if you cook

• When I first saw long, thin eggplants in the late summer markets here on the West Coast, I thought they were some kind of curiosity, a mutation of the full, round eggplant. My wife, who is Japanese, recognized them as being special treats of early harvest time in Japan. She prepared some for me on our charcoal grill. Her method was as simple as one can get. She turned them over the hot coals, cooking them in their skins. When hot throughout and slightly soft, we dipped them in a mixture of soy sauce and freshly grated ginger. I am not a vegetarian, but I've never enjoyed a charcoal grilled steak more than I did those first charbroiled eggplants . . . I've never missed having them among my summer vegetables in 12 years.

—*John Meeker*
Gilroy, California

• It was always a mystery to me why the price of eggplants always jumps in Japan during July and August when eggplants are at their most abundant. The reason, as I later was told, is because during the dry season of early summer, the farmers must mulch and irrigate. To compensate for the extra cost of production the price goes up. Water for irrigation is diverted from the canals which carry water to the rice fields. In upland areas where water is scarce, it is carried in buckets dangling from the ends of a long shoulder pole from streams, ponds and springs.

—Larry Korn
San Francisco, California

the eggplants in a favorite dish and freeze them in that form.

Why grow this vegetable that seems fussy and difficult to store? Very simply, because it is small and tender and its flavor is outstanding.

Because of the vegetable's thin skin, Japanese eggplant is easier to cook than the larger global types, especially if you're fond of oriental methods. The versatile *nasu* can be stir-fried, sauteed, broiled, baked, steamed alone or with other vegetables, added to soups, dipped in tempura batter and deep-fried or even cut into thick slices and grilled. For a Japanese-style treat, the eggplant can be fried or pickled in *miso*. To pickle it, simply wash the fruit, bury it in a crock of *miso,* and in two or three months the eggplant will be permeated with the salty, enzyme-rich juices of this soy paste. The same process can be used for sliced burdock root as well.

W. Atlee Burpee Co.; J.L. Hudson, Seedsman; Johnny's Selected Seeds; Le Jardin du Gourmet; Nichols Garden Nursery; Geo. W. Park Seed Co.; Redwood City Seed Co.; and numerous other seed companies carry the Japanese eggplant under various names. It is important to read the catalog descriptions closely as sometimes they are listed as a Japanese variety and sometimes not. A fast-maturing Japanese eggplant named NAGAOKA NEW KISSIN can be purchased from the Kitazawa Seed Co., a small West Coast outlet that specializes in Japanese vegetables.

Florence Fennel

Foeniculum vulgaris, var. *dulce**

FENNEL is one of those workhorse plants that men have held in high regard from time immemorial as food, flavoring, tea, medicine, and insect repellent. Yet strangely, in modern times—except among people of Mediterranean heritage—it has fallen out of general use. Undoubtedly fennel deserves a better fate, for it is certainly of the easiest culture, and will grow almost anywhere a garden can be established.

Fennel is apparently a native of southern Europe, but it has been naturalized in many places around the world. The ancient Greeks believed it imparted courage and long life to its users, and discovered that by eating the feathery leaves or drinking a strong tea made from them, they could temporarily cut down their appetites. Fennel became one of the first reducing aids.

There are three kinds of edible fennel, one an herb and two vegetables: Common fennel (*Foeniculum vulgaris*), grown for its seeds and leaves that are used to flavor soups and fish sauces; carosella or Sicilian fennel (*F. vulgaris,* var. *azoricum*), grown in southern Italy for its tender young stems that are eaten like celery or asparagus; and Florence fennel or finocchio, which is cultivated for its very thick, basal leaf stalks.

*The botanical name of Florence fennel is listed in some references as *F. dulce;* in others as *F. azoricum.* Its cousin carosella is listed variously as *F. azoricum* and *F. piperitum.*

Florence fennel is a cool-season vegetable that can be grown as an annual throughout this country and as a perennial in the milder regions. Botanically, it is a member of the Umbellifer family and, as such, is related to celery, carrots, and parsley. The plant resembles celery except that its leaves are finer and more feathery. It grows to a height of 1 or 2 feet with three swollen leaf bases that overlap to form a sort of false bulb.

Today's Italians, like their Roman forbears, use all parts of the Florence fennel plant—the bulbous stems, leaves, and seeds. Besides appreciating its food value, the Romans believed fennel improved eyesight. The Germans, French, and English shared similar beliefs, and the early Anglo-Saxons used it in all manner of mystical potions to improve crop yields on the one hand, and to prevent various physical and spiritual maladies on the other. Some modern herbal practitioners affirm that fennel does indeed have specific and valuable medicinal uses as a stomach and intestinal remedy to relieve abdominal cramps and flatulence, and as an eyewash.

Florence fennel requires 80 to 110 days from date of seeding to reach maturity. Like Chinese cabbage, it prefers cool, moist weather and will bolt in hot temperatures. In areas where summers are very warm, it is grown as an autumn or winter crop sown in late summer. In very mild regions, finocchio can be started in autumn or even winter for spring harvesting. Gardeners in the North should plant seeds in May and harvest in summer.

For a good crop of finocchio, the soil pH should be nearly neutral. If your soil tends to be acid, add dolomite limestone as necessary. Adding compost or well-rotted manure to a depth of about an inch will further encourage good growth. When planning the amount of space to be set aside for Florence fennel, figure on 18 inches between the rows for bulb production, and 2 feet if you want leaves for tea and seeds to use for a condiment or medicine.

Sow seeds thinly, ½ inch deep. When the plants have become established, thin them to stand at 6- to 8-inch intervals in the rows. As a precaution against bolting, supply the plants with adequate water throughout the growing season, and cultivate frequently, to a shallow depth, to control weeds.

A good rule of thumb for coping with local climatic conditions is to grow Florence fennel, which is sometimes called Italian celery, as you would grow celery. Fennel is easier to grow, though, because it doesn't need continuous applications of moisture. It is quite able to stand dry conditions, though for tenderest eating you will need to water the plants whenever the top inch of soil dries out.

As the fennel grows, the bulb-like stems begin to form just above the ground. When the plants are about 1 foot high and the leaf bases about 1 inch thick, they are blanched like celery or asparagus. Mound the soil around the leaf base and mulch from the center of the row, around, and covering the "bulbs." The object is to prevent light from striking the stem bases without hindering the growth of upper stems and leaves. Unblanched, the stems would have an unpleasantly strong licorice flavor.

The "bulb" is ready to harvest when it is 2½ to 3 inches in diameter; beyond that stage it becomes tough and stringy. It may be dug or pulled, then the roots and faded greens are removed. Early frosts may damage fennel so care must be taken to make summer plantings early, before the end of July in most regions.

Harvested and washed, Florence fennel can be chopped up raw for salads, braised or stir-fried in butter, or steamed and served with a cream sauce. Its sweet, slightly anise flavor is a nice complement to various cheeses and, boiled with meat or served with macaroni, it makes a terrific, unusual dish. Finocchio is also especially delicious when parboiled, then baked with a sprinkling of melted butter and grated Parmesan cheese. English cooks in centuries past made a cold fennel soup with the sliced stalks, some wine, sugar, ginger, and almonds. They also liked to serve it with fish. Today, you're likely to find Florence fennel on the antipasto platter in restaurants where northern Italian cuisine is featured. Basically, the leaf stalks can be used in all the ways you use celery, with the advantages that the plant is easier to grow, and the vegetable's flavor is more distinctive.

Fennel Hay

When you harvest your crop of finocchio, don't throw away the tops. If you don't use them in the kitchen, spread them out and let them dry thoroughly. You can use this hay as part of the bedding material for cats, dogs, and other domestic animals—it repels fleas. Your pets will appreciate it!

The fresh leaves can also be harvested for use in salads or wherever a slight anise flavoring is desired. The leaves can be used in salad dressings, too, in either fresh or dried form. To make fennel tea, harvest the tops and let them dry in a warm place away from direct sunlight. When the leaves crumble easily, they can be stored in an airtight container. The seeds, which appear the second year if the plant is not harvested, are useful in medicinal teas or as a natural flavoring in cookies and candies.

Many companies carry seeds of Florence fennel or the similar carosella. Suppliers include: W. Atlee Burpee Co.; Casa Yerba; Comstock, Ferre and Co.; William Dam Seeds; J.A. Demonchaux Co.; DeGiorgi Co.; Gurney Seed and Nursery Co.; Joseph Harris Co., Inc. Charles C. Hart Seed Co.; J.L. Hudson, Seedsman; Meadowbrook Herb Garden; Mellinger's, Inc.; Nichols Garden Nursery; L.L. Olds Seed Co.; Redwood City Seed Co.; Seedway; Stokes Seeds, Inc.; Thompson and Morgan, Inc. and Otis S. Twilley Seed Co. Plants can be purchased from Well-Sweep Herb Farm.

• The seeds of Florence fennel, or the oil expressed from them, have long been used in herbal remedies for assorted ailments. Here are just two medicinal uses for Florence fennel:

To make a soothing eye-wash, boil a teaspoon of seeds in a cup of water, strain, and use to rinse the eyes three times a day.

To make fennel cough medicine, mix a few drops of fennel oil with a tablespoon of honey, and take a teaspoonful at a time, as needed.

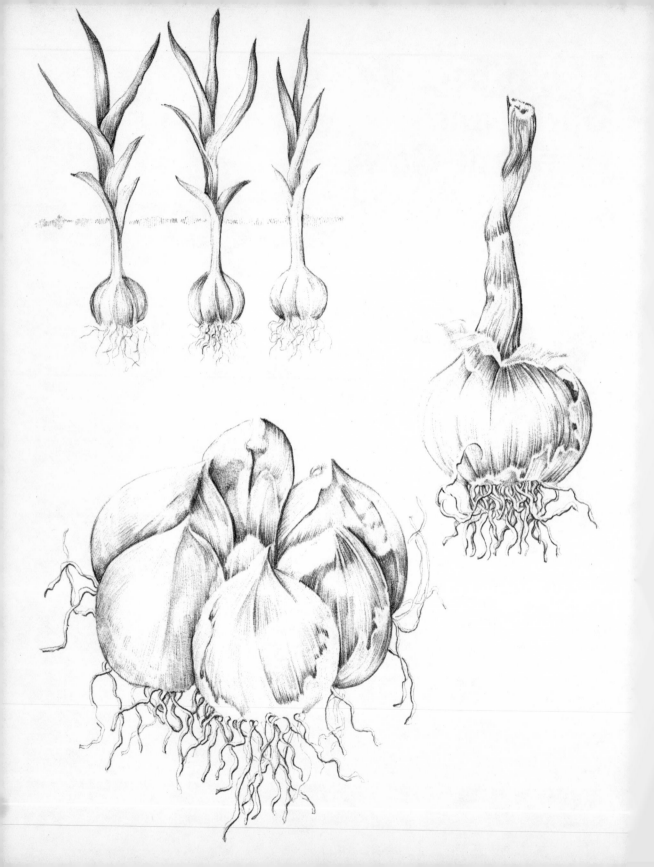

Garlic and Elephant Garlic

Allium sativum and scorodoprasum

THIS hardy perennial has been found growing wild in only one restricted part of the world, the mountains between Afghanistan, Mongolia, and Uzbekistan. It has been locally naturalized elsewhere, and today is cultivated in many parts of the world.

Many ancient cultures were familiar with the culinary and medicinal uses of garlic. It is mentioned in Sanskrit writings and in the book of Numbers in the Bible. Herodotus writes that the Egyptians ate it and that it was a part of the diet of the pyramid builders. Among many peoples, though, the plant was considered an unclean and evil food of barbarians. Ancient Greek priests permitted no one who had eaten garlic to enter the Temple of Cybele. The Romans detested the strong odor, but nevertheless fed the bulbs to their soldiers and laborers. Perhaps they, like the later herbalists, recognized some curative powers in the plant. But nowhere, as a food or medicine, did garlic reach the popularity it enjoys in southern Europe and Asia. To this day the pungent bulbs remain a key ingredient in practically every dish of southern France, Spain, Italy, and central Asia.

There is only one species of true garlic, *Allium sativum,* an herbaceous biennial belonging to the lily family. Its leaves grow 8 to 10 inches long, and its flower stalks are very similar to those of the onion, bearing both seeds and bulblets. Seed, however, is

• Garlic doesn't produce as well for us as it would in a cooler climate; but we get enough for our needs. We love Italian sauces laced with garlic. The biggest problem I have with garlic is grass. One year I neglected the garlic and when I went to harvest, I couldn't even find the garlic among the grass! Garlic definitely prefers the sandy soil of our "North 140."

—*Nancy Pierson Farris*
Estill, South Carolina

185

seldom used for propagation since the bulblets and, more commonly, the cloves which make up the mature bulbs are more easily handled. Unlike its cousin, the onion, garlic produces a compound bulb which consists of up to ten smaller cloves. Each clove is formed from two leaves. Each clove is inside a protective sheath, and the whole compound group is covered with a thin skin. In late-maturing varieties, this sheath is pinkish or brown; early varieties have tan-colored cloves. Late garlic matures two or three weeks after the early type, and produces lower yields, but it stores better and is considered by many cooks to be of higher quality.

The planting and culture of garlic differs very little from that of onions, although some gardeners find garlic more exacting in its requirements. It thrives best in a fertile, well-drained loam, but almost any soil which supports onions will produce a satisfactory garden crop. An open, sunny location which is not too dry in summer is ideal.

Garlic can be planted early in the spring, during the summer, or in the autumn, depending on the climate and day length of your area. Because warm temperatures and long days favor bulb development, sets should be planted early enough to allow the leaves to fully develop during cooler, shorter days. The plants' yield depends on the amount of vegetative growth made before the bulbs begin to form. In mild regions, garlic is best planted in the fall; it will shoot up very early in the spring and easily surpass any sets put out that season.

Since the plants require very little space, they can be grown in borders or interplanted with other crops or flowers. Perennial beds are easily established by simply leaving about one quarter of the crop in the ground to sprout next season.

Plant individual cloves root end down, and cover with 2 inches of rich, sandy soil. If 6 inches are allowed between sets, about ½ pound of sets will plant 100 feet of row.

Garlic is a cold-hardy plant, unharmed by frost or even light freezing. If planted six weeks before the last frost, it will sprout quickly and grow vigorously until midsummer. It is sensitive to frequent rains and should be mulched to ensure a continuous supply of moisture. Keep the garden weed-free or the develop-

• Garlic among roses brings a number of benefits to the flowering shrub. According to those who have studied how plants exude an essence or essential oil into the air, soil and water in their environment, the two make ideal companions. Many rose growers report that garlic planted with their bushes has prevented insects. It can also be an ally in efforts to banish such rose problems as rust, mildew, and blackspot, besides mites, thrips, or any of the beetles which seem to favor feeding on roses.

—Mayme Bobbitt
Jonesboro, Arkansas

ing plants will be overrun. By early June, flower stalks may appear and should be cut back and discarded so that the plants' energies can be directed toward root and bulb development. During the warm months, bulbs begin to mature and ripen, and the leaves wither and yellow. If by midsummer the tops have not fallen, they should be bent. When the leaves have yellowed, lift the plants and dry the bulbs in a partly shady place. Or, braid or tie the tops together and hang in a cool, well-ventilated spot. Garlic bulbs will often keep for over a year if properly stored, in wooden trays, or hanging in bags or bunches. Dampness is an open invitation to rotting.

Most cooks would find it difficult to concoct at least some of their favorite dishes without the versatile garlic. Garlic is an important part of many oriental stir-fry dishes; white clam sauce for spaghetti would be nothing without it; Mexican refried beans taste best when cooked with liberal amounts of it; Middle Eastern broiled chicken with lemon and garlic depends on it. The pungent cloves can be chopped and added to practically any meat or fish dish, as well as salads, sauces, and soups. True garlic lovers even enjoy it steamed whole with other vegetables.

Many gardeners enjoy eating the green shoots and leaves of garlic plants, like chives or green onion tops. Cutting them continuously inhibits bulb formation, but these growers don't seem to mind. They find garlic greens sweeter-tasting and subtler than the cloves, and use them to enhance the flavor of various dishes. As with chives, the supply of garlic greens is an endless one; they can be cut almost all year long.

As an herbal medicine, garlic is said to prevent asthma attacks and relieve head colds. When eaten in liberal quantities it is also reputed to prevent high blood pressure.

Garlic has a number of nonfood uses as well. A garlic-containing oil has been used to destroy mosquito infestations by spraying on breeding ponds. Planted in a circle around fruit trees, garlic is said to keep borers away; planted near tomatoes, it guards against red spider mites. Mixed in a blender with water, it can be sprayed on leafy crops to keep away bugs. The mature cloves help keep weevils out of stored grain when some are kept in the storage container.

• Several years ago we accidentally discovered a way to keep birds away from our corn seedlings . . . Early in the year we decided to plant some garlic in the corn patch as a sort of pre-crop. But no sooner would the garlic sprout than we would find it lying on top of the ground. We would replant it, only to have the same thing happen. But after several days, the birds tired of their game.

When we planted the corn later in the year, the birds left the seedlings alone. Apparently, they thought it was just more garlic which they didn't like. We had a bumper crop of corn from the patch.

—*D.C. Moses*
Huntington, New York

Garlic sets can be obtained from many seed suppliers, among them, W. Atlee Burpee Co.; Comstock, Ferre and Co.; William Dam Seeds; DeGiorgi Co.; Farmer Seed and Nursery Co.; Gurney Seed and Nursery Co.; Joseph Harris Co., Inc.; Jackson and Perkins Co.; Le Jardin du Gourmet; J.W. Jung Seed Co.; L.L. Olds Seed Co.; R.H. Shumway Seedsman; Spring Hill Nurseries, and Thompson and Morgan, Inc. You can also simply plant cloves from the grocery store.

ELEPHANT GARLIC (*Allium scorodoprasum*)

Cloves of this jumbo garlic grow to the size of pullet eggs or small tulip bulbs. Some get even larger. Their large size makes them much simpler to peel and prepare than ordinary garlic; one clove equals about 18 cloves of true garlic. In addition, elephant garlic has a much more delicate flavor than *Allium sativum* and, for those who have never been able to face garlic in the past, it is a welcome addition to the diet.

Growing and cultural requirements for elephant garlic differ only slightly from those of other alliums. For best results, prepare a rich, deeply cultivated bed for the plants. Open a trench about 6 inches deep and place a 4-inch layer of compost, manure or leaf mold on the bottom. Then add 2 inches of topsoil and water well. After the bed has settled, plant cloves, blunt or root-end down, 12 inches apart. A shallow planting depth of about 1 inch produces larger bulbs than does deeper planting.

Mulch the plants after they have become established and keep the bed fairly weed-free. Deep cultivation could damage the bulbs and cause them to rot, so hand-pulling is best. Plants grow to a height of 30 to 36 inches and look very much like underdeveloped corn. In early or midsummer, after the eighth leaf appears, a long flower stalk arises from the center of each plant. If left to develop, it will produce a globe of violet flowers. As with true garlic, these flowers should be cut back as they appear so that the plant's energies are not diverted from bulb formation. As the bulb matures, the plant's lower leaves will

• A planting of elephant garlic among my melons and cucumbers seemed to have a very good effect in deterring insect damage. Whenever I crushed a leaf tip the air was immediately filled with the garlic odor to which the insects seemed to object. I noted that where it was growing close to a tomato plant, that plant would be free of the ugly green tomato worms which I picked off other tomatoes not near these sentinels.

—John Krill
North Lima, Ohio

gradually yellow and fade. At this time, water should be withheld. When the top has fallen over, growth is complete and the bulbs can be lifted. Allow the bulbs to dry in the shade for three or four days, then shake the dirt from them and store in a cool, dry place. The little corms that have formed at the base of each bulb can be saved and planted the following spring. Each will produce a small, onion-like bulb which is saved the following winter and then grown yet another season to produce a compound bulb.

Elephant garlic is mild and sweet enough to be served steamed as a vegetable dish. It's especially tasty with a cream sauce. It can be sliced raw and added to salads, or cooked and served, like onions, with a pot roast. It can also be substituted for true garlic in most dishes, when a less pungent flavor is desired.

Seed bulbs for starting elephant garlic are available from Burgess Seed and Plant Co.; Casa Yerba; Farmer Seed and Nursery Co.; Gurney Seed and Nursery Co.; Lakeland Nurseries; Nichols Garden Nursery; R.H. Shumway Seedsman; and Thompson and Morgan, Inc.

Good King Henry

Chenopodium bonus henricus

GOOD King Henry is an example of the many perennial potherbs once widely cultivated in European and colonial gardens and now flourishing in the wild on both continents. With the perfection of beet, chard, and spinach, this British native lost favor to the extent that today it is only rarely cultivated in European gardens and all but forgotten in this country. It is, however, a very versatile and wholesome early spring vegetable and an excellent choice for the gardener with plenty of space but little time for cultivation. With a minimum of care, it produces bountiful harvests of food, fodder, and compost material.

A member of the goosefoot family, it is a coarse, many-branched plant which grows to a height of about 2 feet. Its leaves are smooth, shaped like arrowheads, and colored dark green with purplish centers. Young leaves are cooked and eaten like spinach, and the shoots and flower heads are prepared like asparagus. The common names wild spinach and Lincolnshire asparagus refer to these culinary uses, while another alias, fat hen, can be explained as a reference to the older leaves that are used in fattening poultry. The plant's medicinal properties were also held in high regard in times past, and it was widely known as Mercury goosefoot and wild Mercury, in honor of the ancient Roman god of medicine. The plant's versatility gave it yet another local name, all-good (*toute bonne* and *all gut*) and probably led to the most familiar name, Good King Henry, which in England refers to the much-loved king Henry VII, and on the Continent to the Dutch and German phrases, *gulden Henrich* and *guten Heinrich,* meaning "good fellow."

Seeds of this plant may sometimes be purchased from small, local nursery operations, or from herb gardens. But you can also start your crop with some transplanted wild plants or get seeds or starts from someone already cultivating Good King Henry. The plant is fairly well-naturalized throughout the eastern and central United States where it is often regarded as a weed. Gardeners and farmers who don't eat it will be happy to have you dig some. In the western states it is locally available from gardeners who have established it from the East.

The plant will do well almost anywhere. It thrives in odd, sunless corners or bits of neglected land. But for the best results, plant it in a sunny location where the soil is rich and well-drained. If the garden bed has been prepared the previous fall, seeds can be sown in early spring as soon as the ground can be worked. Sow seeds very thinly, ⅛ inch deep in rows about 18 inches apart. When they are large enough to handle, thin seedlings to stand 1 foot apart. Young plants transplant very well and root easily. If seeds are unavailable, Good King Henry can also be propagated by root division.

Keep the bed moist and free of weeds throughout the summer. Once established, the plants will continue to grow and self-sow for years, though the bed should be cleaned out and a new crop started after 10 years. Left to their own devices, the plants will seek out the richest garden soil, where they will form deep rootstocks and send up much coarse growth. They are not exactly pretty plants, but with the flashes of red against bright green leaves, they are certainly not dull. If the rank growth is controlled and the many new seedlings are weeded out each spring, the Good King Henry can be put to work occupying spaces where less desirable weeds would normally take over.

Although many gardeners eat and enjoy the foliage all summer long, most prefer to use the plant when it is "in season"

• The nurseryman who sold me the seeds of Good King Henry advised me to cover the bed with straw after the first freeze to insure an earlier spring crop. I found that I could have some even earlier by putting a bottomless box over a corner of the bed, and covering it with a pane of glass, thus forcing a few plants into very early growth. The leaves when young and tender are excellent eating, a good substitute for spinach. And the flower buds, cut and steamed, are every bit as good as asparagus and available much earlier.

—*Dorothy L. Baker*
Duluth, Minnesota

from April through June. Cut some of the young stalks as they appear in early spring. If the leaves were cut down the previous fall and the plants heavily mulched, this new growth will appear very early, in March or April. A hot cap placed over the plants will further speed their development and make Good King Henry the first fresh green vegetable of the season. Young stalks need no preparation other than cleaning before they are steamed and eaten like asparagus, which they resemble in flavor. Try them with hollandaise sauce, lemon butter, or herb butter. Older stalks should be peeled to remove the tough, outer skin before cooking.

Later, the young leaves can be harvested and prepared like spinach or other greens. Steamed in only a small amount of water, these greens are a delicious source of vitamin A and several minerals. They're quite good creamed and topped with chopped hard-boiled eggs, prepared au gratin, or cooked with mushrooms. It is important that only the young leaves be used, as the older ones tend to be very bitter, requiring two changes of cooking water to make them palatable. This seems to be a terrible waste of vitamins and minerals and, unless you're desperate for greens and have no spinach, chard, or New Zealand spinach to use, these older leaves are best avoided altogether. Once the plants have gone to seed, they are all but inedible. They are best used as animal fodder or compost material.

Nevertheless, the very industrious gardener-cook can continue to make use of Good King Henry. Although it is a time-consuming task, and not really economically feasible, mature seeds can be gathered and ground into an acceptable meal.

If you're not inclined toward this sort of effort, pull up the plants and toss them on the compost heap, or feed them to the livestock.

Good King Henry plants are available from Well-Sweep Herb Farm, and seed can be had from Casa Yerba.

Ground Cherry

Physalis pruinosa

MOST of us assume that names were distributed to everything and everybody to aid identification and eliminate confusion. However, this isn't always the case. It certainly isn't the
case for the ground cherry, which seems to break all the rules of
nomenclature. This little berry is not a cherry; it is sometimes
referred to as the strawberry tomato, even though it's neither a
strawberry nor a tomato. Then again, some people call the
ground cherry a husk tomato, which is not to be confused with
the cherry tomato, a true tomato about the size of a large
marble. To further complicate matters, in certain parts of the
country this herbaceous annual is called the dwarf cape
gooseberry.

Whatever name you decide on, the ground cherry is one of
several species belonging to the genus Physalis in the nightshade family. The plant produces a small, round yellow fruit
housed inside a thin, papery husk which resembles its close
relative, the Chinese lantern plant so well-known as an ornamental. The fruit is the edible portion of the plant and its
sweet-tart, slightly acidic flavor is delightful in pies and preserves.

Many species of the edible ground cherry are native to
Central and South America. Those with good eating qualities
have been introduced to the United States where they now can

Ground Cherry Jam

Here's a dandy ground cherry preserve we like. You'll need 2 pounds of ground cherries (about 8 cups husked), 4 cups sugar, 1 cup water, and the grated rind and juice of 2 lemons. Husk and wash the ground cherries carefully. Measure the sugar and water into large kettle. Bring to a full rolling boil, and boil for 2 minutes.

Add the cherries, lemon rinds, and juice. Bring to a full rolling boil again, reduce heat and simmer for 5 minutes. Remove from heat, cover with a clean towel, and let stand overnight. Next day, return to the heat, and again bring to boil. Reduce heat and cook gently until transparent (about 15 minutes). Immediately pour into hot, sterilized glasses; seal at once. Yields 5 to 6 cups.

—Doc and Katy Abraham
Naples, New York

be found from New England to the tip of Texas. Many of the wild ground cherries that have spread throughout this country, including Hawaii, are direct descendants of plants that escaped from gardens. Only rarely is the ground cherry included in today's American gardens, an unfortunate state of affairs in light of its countless uses, ease of cultivation and notable productivity.

Growing habits and general culture of the ground cherry are very similar to those of its cousin, the tomato. Ground cherries prefer a sunny, well-drained location. Most species need a fairly long, warm growing season (about three months), so in northern areas, seeds should be started indoors or in a greenhouse. Start the seed in peat pots in a mixture of one part each of sand, peat and loam. As soon as growth begins, gradually harden off the seedlings. After six weeks transplant them to the garden. For best results, plan the date of transplant to occur approximately five days after the average date of last spring frost. Space the ground cherry plants about 3 feet from each other, cultivate frequently, and water the ground around the roots of the plants when it dries out. Keep soil loosened between the rows by frequent hoeing.

Ground cherries also grow nicely in the greenhouse. One good method is to start the seed in peat pots, as described above. When the plants are 2 inches tall, they may be planted in 10-inch pots, or directly in the bench. Space the plants at least 18 inches apart, and be sure the soil is well drained. Ground cherries in the greenhouse are susceptible to white flies, which can be controlled with a small parasite called *Encorsia fermosa,* or anything with nicotine in it. Tobacco dust or a tobacco tea thickened with soap can be applied to the undersides of the leaves to get rid of the white fly. Lisa Widman of Brooklyn, New York, also reports that she has successfully grown husk tomatoes indoors in hanging baskets. The vines grow about 1 to 2 feet long, and trail nicely from the pots.

Growth in the garden is low, vining and very full. One California gardener planted his first ground cherry seeds directly into the ground between rows of half-grown cabbages and broccoli, covering the seeds with ½ inch of well-prepared soil. After thinning he watched the plants grow . . . and grow and grow. "The plants were huge," he said. "They never grew more than a foot or so high, but they rapidly spread in every direction, climbing over low obstructions with ease. My original idea was to use them as an interplant but unless they grow between tall, widely spaced plants like hill-planted corn, they'll overrun their neighbors. Happily, I harvested the cole crops before there was any serious competition for light and nutrients."

The fruits are mature when they turn yellow and somewhat soft. They will fall off the vine when they're dead-ripe. Try to pick off the loose fruit at least once a week. Their husks help prevent rotting, so ripe ground cherries won't deteriorate for several days even if left on the ground. The average yield is about 2½ pounds of cherries per plant.

If you gather more ground cherries than you can use at once, spread them out in a cool place and let the husks dry. You can then store the ground cherries for several weeks in a cool, dry place. Some people report good luck in storing ground cherries, but most seem to feel the fruit is not adapted for long-term storage, and should be used or processed not too long after the harvest.

Although the fruit may be eaten raw, it is usually boiled and used in stews, dessert sauces, preserves, ice cream toppings, salads and pies. When dried in sugar ground cherries make an excellent fruit to use in fruit cakes; it's a lot cheaper and, in the opinion of many cooks, tastes better than citron, figs, or even raisins.

Seed houses that carry ground cherries (alias husk tomato, strawberry tomato, or cape gooseberry) include Burgess Seed and Plant Co.; DeGiorgi Co.; Henry Field Seed and Nursery Co.; J.L. Hudson, Seedsman; Earl May Seed and Nursery Co.; Nichols Garden Nursery; L.L. Olds Seed Co.; R.H. Shumway Seedsman; and Thompson and Morgan, Inc.

197
Ground Cherry

Saving Ground
Cherry Seed

The second year I grew ground cherries, I realized I could get a jump on the season by starting seed indoors . . . handling them the same as I would tomatoes. As it turned out, the extra work wasn't necessary. When chilly rains and cold destroyed the plants in mid-fall I allowed the remains to decompose on the spot. In the spring, after the ground warmed up, volunteer plants sprang up everywhere the vines had been. Some I allowed to remain where they sprouted. These produced before the indoor-started seedlings. Others I transplanted to various locations in the garden, and they came in about the same time as the indoor-started sprouts. That was my last year of buying seed, and the first year of selecting the plants with the best-tasting berries to provide volunteer stock for the next season.

—*James Jankowiak*
Eureka, California

Horseradish

Armoracia rusticana

HORSERADISH is indigenous to eastern Europe and the Middle East but has become naturalized in many north-temperate regions of the world. Its history as an herb and food is a long one. It is said to have been one of the bitter herbs eaten by the Jews during Passover. Its medicinal properties were well-known in ancient times and, beginning in the Middle Ages, its leaves and roots were used in Europe as food. Brought to the United States in the seventeenth century, it soon became a standard crop in northern kitchen gardens.

Horseradish is a hardy, perennial member of the Crucifer, or mustard family. Several varieties are available, each of equal quality. Among the most widely grown are NEW BOHEMIA and MALINER KREN. All horseradishes form stout, yellowish to white taproots. Leaves develop in a large tuft from which a stalk arises bearing small, white flowers. Seed is produced, but it rarely matures and is never used in propagation.

Instead, horseradish is propagated from crown or root cuttings planted early in the spring. For regions with short growing seasons, the crown method is generally considered best. By this technique, a horseradish root with crown attached is dug in early spring or late fall. The piece is split lengthwise into as many strips as possible, each strip consisting of both crown and root. Let the cut wounds heal for several days, then,

> To make a good cold horse-radish sauce that's a delicious accompaniment to cold ham or roast beef, just whip ½ cup of heavy cream until stiff and fold in 2 to 4 tablespoons of grated horseradish.

199

• Horseradish is an excellent mucus eliminator—a natural way to open up clogged sinus passages and get them to drain freely. Try mixing it with honey. Use one tablespoon of honey with ½ to 1 teaspoon of ground horseradish. Cover and let marinate a few hours. Eat slowly and the fumes will reach the sinus areas to break up the phlegm and congestion of a cold. Tongue or throat coating often disappears by morning, too. It's better to grind horseradish out-of-doors since the fumes released open tear ducts and produce "horseradish tears" and sinus drip whether congested or not.

Learning how to eat the horseradish and honey preparation takes some practice and patience. For anyone just acquiring the knack, diluting it with ground turnips and blending with vinegar or lemon cuts the strength to more palatability.

—*Elsie V. Alfrey*
Rockford, Illinois

using a dibble, make holes in the soil and plant each strip at a slight angle, dropping the tops 1 or 2 inches below the soil surface. Firm the ground around the root and water well.

The best place to get crown cuttings is from a neighbor's horseradish bed; most nursery seedsmen supply only the ordinary root cuttings that have been trimmed from the larger, marketable roots. They range from ¼ to ½ inch in diameter and from 5 to 10 inches long, the longer cuttings being preferable. The top end of a root cutting is always trimmed square, the other end sloping, to show it is to point downward in the ground. Be careful to set them in the ground the right way; if you plant the cuttings upside down the crop will probably fail. The cuttings are planted 4 inches deep in rows about 2 feet apart. Lay the roots at 1-foot intervals in the furrows. The square-cut tops should stand slightly higher than the bottoms and should point in the same direction. Firm the soil over the sets (but do not pack) and again, water liberally. Root cuttings are a good way to propagate if you want to get a lot of starts from a small number of plants.

The ideal soil for horseradish is a deep, moist loam with underlying clay. A high organic matter content is crucial and, because the roots grow at least 10 inches in length, an abundance of potash is required. Hillsides and bottomlands are excellent, but composting, mulching, and other humus-building practices will make all but the sandiest or heaviest soils suitable for horseradish. The ground should be prepared deeply and allowed to settle several days before planting. Manuring is best done sometime before planting, in order to avoid too much top growth or irregularly branched roots. Although horseradish will grow anywhere and indeed is hard to eliminate from old gardens, the roots are much larger, tastier, and freer of branches when given proper care and cultivation. Also, it's a good idea to keep horseradish off ground that has been planted to other members of the cabbage family in the past three to five years, to avoid building up localized concentrations of the pests and diseases that mutually attack these vegetables.

Hoeing between the plants is advised during very early

growth but other than this, little care is required. In order to ensure smooth, strong roots, some gardeners dig up the roots when growth starts and remove all shoots except the one at or nearest the top. Market gardeners go a step further to insure straight roots, uncovering the upper parts of the roots when the plants are 9 to 12 inches tall, rubbing off all the lateral roots, then replacing the soil. This process may be repeated several times during the season provided care is taken that the bottom, feeding roots are not disturbed. The resulting taproot is long, straight, and white. It is very marketable, but it doesn't taste any different from those "freeform" roots allowed to grow undisturbed.

Although horseradish is a perennial, the common practice is to harvest the roots every autumn and replant the starts in the spring. Even if your horseradish is not grown exclusively as an annual, the beds should be moved every two or three years. After that length of time the roots begin to divide and harden.

Horseradish needs a climate where autumn and winter weather is fairly cold. It does not acquire its characteristic thick, pungent taproots until cold weather has encouraged the plant to begin storing starch. Roots dug in summer are small, weak-tasting, miserable specimens; the best roots are those that have endured several frosts. Of course, they can be left in the ground all winter, but it is better to dig them in autumn and store them so that they will be available as needed. To harvest, dig up the roots and remove the tops and side shoots. Since horseradish has a tendency to overrun the garden, it is important to remove all the roots at harvest time.

The roots can be stored in moist sand or sawdust in a cool, dark root cellar. Or, place them, sand and all, in a plastic bag and store in the refrigerator.

For an interesting addition to winter salads, the roots can be sprouted by placing them, with the crowns, in moist soil in a dark, warm place. The leaves that develop will be white, tender, and sweet. Cut the leaves when they are 3 or 4 inches long to use in salads. Be sure you grow these leaves in the dark, though, or you will find them tough, strong-tasting and green.

The roots themselves are useful both as food and

• Here's an old recipe for that English favorite, horseradish sauce. The recipe comes from *The Book of Rarer Vegetables*, by George Wythes and Harry Roberts, published in 1906 by John Lane Company.

Simmer a small teacupful of finely rasped Horse Radish in half a pint of broth. Then thicken, by adding the yolks of two eggs beaten up with a dessert-spoonful of tarragon vinegar. Add a little pepper and salt and serve. To make cold Horse Radish sauce, add to a small teacupful of gratings, half a pint of mayonnaise. Serve as cold as possible.

• I'm indebted to a college roommate of Japanese heritage for an external cure for rheumatism or muscle aches: Grate the root into a thick cloth, spread it out, and apply the cloth next to the skin. When you feel the heat, remove. This is rather like the old-fashioned mustard plaster, and equally efficacious if you have the fresh root to spare.

—*James Jankowiak*
Eureka, California

medicine. Stalwart Germans sometimes clean, slice, and cook the whole roots the way most of us less hardy souls prepare parsnips. Grated and mixed with vinegar and spices, they have a stimulating, pungent flavor, and are delicious as a condiment with practically any dish. Horseradish is often added to tomato catsup to make a hot seafood sauce. A 2- or 3-inch section can be added to cucumbers and pickled in a brine of cider, salt, mustard and raw sugar. A small amount will spice up pot roast or baked ham, and a few drops of the juice gives a surprisingly different flavor to cole slaw, applesauce, and cottage cheese.

The fresh root can be grated and stored in vinegar for use, by the teaspoonful, in curing rheumatism. Mixed with honey, horseradish will ease coughs and the congestion of a cold.

Once grated, horseradish should be used or stored in vinegar right away. The root loses its pungency when it's exposed to air and heat.

In spring, the first leaf shoots of the horseradish plant can be picked for an unusual and pungent potherb.

Although the English essayist William Cobbett recommended an 8-by-16-foot bed for the family "that eats roast beef every day of their lives," most gardeners find that a few plants satisfy all of their culinary and medicinal needs. Horseradish is a good plant for the cold frame, or for a spot under the greenhouse bench where it can receive some light. To save space in the outdoor garden, it can be planted 8 to 10 inches deep between early cabbages. The deep planting retards growth until the cabbage has been harvested. Horseradish will also do well planted in association with potatoes. A few plants placed at the corners of the potato patch will deter blister beetle.

Horseradish roots can be obtained from several growers, among them, Burgess Seed and Plant Co.; W. Atlee Burpee Co.; Casa Yerba; DeGiorgi Co.; Farmer Seed and Nursery Co.; H.G. Hastings Co.; Nichols Garden Nursery; Seedway; R.H. Shumway Seedsman; and Spring Hill Nurseries. J.W. Jung Seed Co. handles crowns, and Well-Sweep Herb Farm has plants.

Jerusalem Artichoke

Helianthus tuberosus

THIS much-neglected brother of the sunflower, known also as the *girasole* or sunchoke, is neither an artichoke nor is it from Jerusalem. A true all-American, this vegetable is a native of North America, and grows everywhere from Nova Scotia to the Mexican border. When Samuel de Champlain and other European explorers arrived in the New World in the early 1600s, they discovered among the foods eaten by the Indian natives on Cape Cod a strange, knobby, white-fleshed tuber. Finding the tubers quite palatable, the explorers sent some back to France, where they soon became popular and were called *pommes de Canada* (Canadian apples), or *batatas de Canada* (Canadian potatoes). Their cultivation soon spread to Italy, where they were grown in the famed Farnese gardens. It was in Italy, also, that part of this vegetable's odd name had its beginnings. The "Jerusalem" part of the name is believed to be an English corruption of the Italian word *girasole,* which means "turning to the sun," a term applied to the new sunflower-like plant. "Artichoke" is believed to have come about because the flavor of the tubers reminded Champlain and his companions of globe artichokes. Most people today don't agree with that judgement, but the name Jerusalem artichoke, however unfitting, has remained.

The Jerusalem artichoke can be grown throughout the United States today, but it does best in the cooler zones. It is a hardy perennial; its root system spreads far and wide from the main stalk, forming tubers which produce new plants the following year. They come in several varieties, but many gardeners regard the AMERICAN strain, which used to be

205

called the IMPROVED MAMMOTH FRENCH, as the best. This type is smoother and more uniform than other strains, and has fewer tiny cracks, making it easier to clean. The *girasoles* which the early explorers found when they came to America were small and red-skinned, but constant experimentation and cross-breeding have developed the white-skinned types of today.

The plants grow like sunflowers, stretching 6 to 12 feet tall. In late summer and early fall, they produce yellow, daisy-like blossoms which a lot of people cut for flower arrangements. About a month after the flowers have faded, when the plants turn brown and die down, the tubers can be dug. Tubers range in size from a few inches across to the size of a man's fist. They are gnarled and knobby, with thin skins and crisp, white flesh. Nothing is wasted when you grow sunchokes. In addition to the pretty flowers and tasty tubers, the stalks and leaves make good forage for animals, or a welcome addition to the compost pile.

Plants are propagated like potatoes, by planting seed tubers. You can plant whole tubers, or cut the tubers into chunks, each of which has an "eye." There are several ways to plant them. One way is to plant them 6 inches deep in hills, one chunk or tuber to a hill. Hills should be about a foot apart in the row, with rows spaced 3 to 4 feet apart. Another method is to simply place the tubers 2 or more inches deep and a foot apart in loose, sandy soil. Jerusalem artichokes are tough—it seems they'll grow no matter how or where you plant them. In fact, if you're not careful, you may get more than you bargained for. Some gardeners have buried the peelings from Jerusalem artichokes along with the rest of their kitchen scraps, only to find a healthy crop of *girasole* plants growing in the same spot the following spring!

The most important thing to remember when planting the tubers is to put them in an out-of-the-way corner of the garden so the large plants don't cast shade over the rest of your vegetables. They can form a nice-looking screen when planted against a fence or in front of a compost pile.

When gardening space is at a premium, you can also grow sunchokes in containers. The plants must be removed and begun again each year, though, for a plant will quickly fill a can

• Farmers in France and England have grown Jerusalem artichokes as food for their animals for hundreds of years. To those who are able to grow *girasole* as a field crop, the practice followed in Great Britain is worth thinking about. There they create a field of *girasole* one year to provide forage for pigs the following year. The year that the pigs forage there accomplishes several things in one operation. The pigs fatten on *girasole* tubers and stalks; in grubbing for the tubers the pigs turn over the soil and uproot other weeds and roots growing in the field; and the dung from the pigs fertilizes the field for crops to come.

with roots and tubers, and exhaust the soil of a small space. Two years of growth will literally burst a wooden tub apart. And any prolonged stay in a container will increase the danger of rot among the tubers. Jerusalem artichokes can also be grown in the greenhouse, but since they're so hardy, you'd be better off reserving greenhouse space for more delicate crops.

For a good return on time and effort invested, *girasoles* are hard to beat. Once in the ground, they practically raise themselves. The young plants usually only need to be weeded once, for they quickly grow bigger and tougher than the weeds. Mulch the plants with grass clippings or other material if you like, but don't cultivate around them or you will run the risk of injuring some of the developing tubers. In California and other areas where summers are extremely dry, mulch is a must, and the plants will probably need watering as well in order to survive.

As for pests and diseases—*girasole* seems to be practically immune. Nothing seems to bother it, except for mice, which have been known to burrow underground and eat the tubers.

To say Jerusalem artichokes are prolific yielders is an understatement. They produce three to four times the harvest of white potatoes; in fact, it's been said that they are capable of producing up to 20 tons per acre! The sunchoke harvest can, at times, be too profuse, so that some tubers are missed during the digging. Not feeling lonely for long, though, the neglected tubers will simply send out shoots and produce lots of company for themselves come spring. Needless to say, such productivity quickly becomes a source of consternation among gardeners who find their plots overrun, and stories like this one aren't at all uncommon:

> Each year for the last ten years since I planted one root of Jerusalem artichoke in a small garden where we like to start peas, spinach and radishes, my husband has dug up all the artichokes he can find and claimed that there, at last, he's got rid of those things.

> Not at all. Back they come in great profusion the next year, and we have a good crop once again.

Rule one for harvesting sunchokes, then, is to take some

• A friend of mine grows her Jerusalem artichokes in her hen yard. As you can imagine, the hens just love it, and nibble the plants right down to the ground. Even so, I've seen my friend dig down into the soil of the yard and bring up plump, crisp tubers, all ready to boil up or bake for supper.

—*Catharine O. Foster*
Bennington, Vermont

pains to get all the tubers if you're not planning on increasing the size of your patch next year. Before frost hardens the ground, you may be able to simply grasp the base of the stalk and pull up the whole plant, roots, tubers, and all. That's the easiest way to harvest, but to the true connoisseur, the tubers aren't worth eating until after they've been hit by frost, when they generally have to be dug from the ground. Charles Wilson, a Vermont gardener, usually finds that it saves him both time and cuts to use a sharp shovel or spading fork to take out the center clusters of tubers under the stalk bases. This way, you can free the clustered tubers of most of the encumbering soil. If you're digging them with a hoe, it's best to begin at least one foot from the planting row to avoid mutilating the tubers that grow on the lateral roots.

Jerusalem artichokes are completely hardy, and they can be left in the ground all winter to be dug as needed, even in the coldest areas. To prepare your crop to winter over in the garden, cut off the stalks about 6 inches above the ground, and cover the plants with a good, thick layer of mulch to keep the ground soft enough to dig through. Be sure you get all the tubers out by spring, though, or they will begin to grow again.

The *girasole*'s great value as a food is that it is completely starchless, storing its carbohydrates in the form of inulin rather than starch, and its sugars as levulose, the way most fruits do. This makes it a great boon to diabetics, who can often use it as a substitute for carbohydrates from other sources in their diets. Dieters appreciate the sunchoke, too, as it has only a tenth the calories of white potatoes. In addition, the tubers contain assorted vitamins and minerals, notably B_1 and potassium.

Once dug, sunchokes are tricky to store because their thin skins allow them to shrivel easily. You may find they'll keep in the crisper drawer of your refrigerator for weeks or even months, but a better place is a box of moist sand, kept in a cool place. They'll survive until spring if they do not get dry. The best storage bin, though, is Mother Earth; it's simplest to just dig the tubers as you need them.

Some folks complain that *girasole* is hard to clean. But when asked how they go about it, they usually admit to attempting to

• Last spring after my garden was planted, I came home from the supermarket with a small bag of Jerusalem artichokes. I had never seen them grow, so I cut a small one in three pieces and planted them at the ends of three rows of vegetables. They all came up and began to get taller. By mid-October they were 10 feet tall. I really didn't expect to see many artichokes when I dug them, but I could hardly believe my eyes. I got a whole bushel from those three plants.

—*Lucille Mogck*
Sioux Falls, South Dakota

peel the tubers like potatoes. This approach is both a great deal of trouble, and unnecessary. All you need to do is wash them under water, brushing off the dirt with a stiff brush. The delicate skin does not need to be removed at all, although some of it will come off in the scrubbing. Cleaning sunchokes well in advance of eating them is a mistake, too. If you have to dig the roots a few days before you are ready to use them, leave them in a plastic bag in the refrigerator, dirt and all, until it's time to prepare them.

Harvested before frost, Jerusalem artichokes have a flavor that is, while not unpleasant, merely bland and uninteresting. But after frost has touched them, they take on a sweet, nutty taste and a crisp texture that adds life to any winter meal. They can be eaten raw by themselves, the same way you'd munch crisp celery or radishes, or they can be added to salads. Sunchokes are delicious added to a potato salad dressed with a tart vinaigrette sauce. You can add them to your favorite vegetable soup or make them into their own cream soup. The tasty tubers can be steamed, sauteed, stir-fried, and cooked in most of the same ways you prepare potatoes. They are especially delectable when parboiled, then sliced and sauteed in butter until golden brown and topped with parsley, chives, rosemary, or your favorite herbs. Or make them into home-fries and have them for breakfast. They also make a tasty sour pickle—a godsend to the gardener with a larger crop than he or she knows what to do with.

If you're slicing the tubers to use raw in a salad, you will need to rub the cut sides with lemon juice, or put the slices in a bowl of cold water to which some lemon juice or vinegar has been added, for they will discolor very quickly, just as apples do when cut.

Seed tubers for Jerusalem artichokes are available from the following suppliers: Burgess Seed and Plant Co.; Casa Yerba; Gurney Seed and Nursery Co.; H.G. Hastings Co.; Le Jardin du Gourmet; Johnny's Selected Seeds; Lakeland Nurseries; Earl May Seed and Nursery Co.; Nichols Garden Nursery; Geo. W. Park Seed Co.; Spring Hill Nurseries; and Thompson and Morgan, Inc. You can order plants for both red- and white-skinned types from Well-Sweep Herb Farm.

Johnny Sunchoke

We live in an extremely dry desert area, with water so hard it leaves a salt deposit everywhere we water, so it is not easy to grow things.

Last year we had found that the Jerusalem artichokes were the only thing that grew beautifully, unassisted by much more than some water. A friend from the mountains near the Apache reservation had given us some wild, Indian artichokes which are bright red-skinned and grow quite as large as our white ones, only somewhat more lumpy. They taste terrific.

Our neighbor children and daughter (all between 7 and 9 years old), and ourselves, with rucksacks on our backs, made a historic Johnny Sunchoke trip one day, and planted the beginnings of sunchoke patches all up and down the Tularosa Creek from our place. We check them now and then, and many have taken hold, getting ready to provide us with nice wild patches of sunchokes. . .

—*Tote Pickering*
Tularosa, New Mexico

Jicama

Pachyrrhizus erosus; P. tuberosus

THIS vegetable is relatively new to California markets and remains virtually unknown to cooks and gardeners in the eastern United States. A brownish root shaped like an irregular turnip, it has a very tough and thick skin that peels off easily, leaving the sweet, white flesh beneath. Roots range in weight from 1 to 6 pounds, and in diameter from 3 to 6 inches. They are best eaten raw, but can also be prepared like sweet potatoes or, when chopped or sliced and added to various oriental dishes, used as a substitute for water chestnuts. The jicama's low starch content and relatively few calories make it an excellent snack food for dieters.

The scientific classification of the vegetable is somewhat confusing and perhaps even uncertain. Jicama (pronounced hee-*kah*-ma) is simply a Spanish name meaning, "edible storage root." Natives of Mexico and South America apply the word broadly to many very different plants, among them several members of the morning glory, dahlia, and bean families. Morning glory jicamas, *Ipomoea bracteata* and *Exogonium jicama*, resemble sweet potatoes in their growth habits. The *Dalia rosea* is a small ancestor of the garden dahlia, grown in Mexico for its edible roots as well as its brilliant flowers. But the correct, "true" jicama sold in the streets in Mexico and available in California vegetable markets is *Pachyrrhizus erosus* or *P. tuberosus*, a leguminous plant. Also known as the yam bean, this vine has heart-shaped leaves and tremendous vines that reach lengths of 20 or 25 feet. The showy flowers are purple or white and very aromatic, but like the seeds they produce, are extremely toxic.

Two varieties of this jicama are available: *jicama de agua* which secretes clear juice when bruised, and *jicama de leche* which contains a milkier liquid. Both are native from Mexico to Argentina and are now widely grown throughout the tropical and subtropical regions of the world.

As a tropical plant, jicama requires a growing season of nine warm months in order to produce large tubers. Southern, coastal regions are ideal, but if a light and very rich soil is provided, a four-month season is often sufficient. Tubers will be small when grown under such conditions, but they will be just as sweet and crisp as the larger ones. Propagation from small tubers produces mature plants more quickly than growing from seed, but such cuttings or tubers are difficult to obtain in this country. Seeds are more commonly used, and by starting them indoors, a fairly good harvest of jicama can be had even as far north as Massachusetts.

Before planting, soak the seeds overnight in warm water. They can be planted the following day or drained and kept in a warm, damp place until they have germinated. Sprouted seeds are more difficult to handle, but seedlings will emerge from the soil within days of planting.

Sow soaked or sprouted seeds indoors in flats, or outdoors as soon as the soil is thoroughly warm and all danger of frost has passed. Like other root crops, jicama demands a great deal of potash and fairly deep cultivation. Loosen the top foot of soil and work in wood ashes and well-rotted manure. If you've started the seed indoors, set out the seedlings at 6- to 8-inch intervals, in rows about a foot apart. To direct-seed, sow thinly 2 inches deep in rows 1 foot apart, then thin the seedlings when they are about 2 inches high.

Water the plants frequently and gently until they have become established. When the plants are large enough, mulch them heavily with clean straw or sawdust and, if necessary, side-dress with compost and wood ashes. Keep the vines pruned to 3 or 5 feet and pinch off the flowers as they appear in midsummer. Such practices will help to direct the plants' energies toward root development.

Very young pods are sometimes picked and cooked like string beans, but the mature beans are extremely toxic and should be kept away from children. These seeds contain

JICAMA STIR-FRY WITH TOFU

Mix a marinade of:

1 cup soy sauce
½ cup water
1 teaspoon honey
1 clove garlic, crushed
½ onion, chopped
½ teaspoon powdered ginger

Pour the marinade over 1 pound of the tofu that's been cut in small cubes. Let soak for an hour; drain, and reserve marinade.

Saute together in a small amount of oil:

1 onion, sliced
1 clove garlic, minced
  the tofu

Add and stir-fry briefly, until hot:

½ pound jicama, thinly sliced
1 carrot, chopped
½ bell pepper, chopped
  Sliced zucchini, yellow squash, or whatever you have available (this is a very flexible recipe)
¾ cup of the marinade, or to taste.

Serve over hot rice, and pass the remaining marinade for seasoning at the table. Serves about 6.

rotenone, which is a potent insecticide and is often used by South American Indians to poison fish.

The rounded, brown-skinned tubers will be formed by late summer. In very warm zones, roots can be harvested up to 18 months after planting, but elsewhere they must be dug before frost. Left in the ground over the winter, they will rot. Dig jicama tubers as soon as the vines have begun to die down in the fall. Like potatoes, they should be dusted off but not washed before being stored. They keep best if refrigerated and better yet if they are peeled before being stored. In cold storage, they will remain fresh for many weeks. Since the tubers do not discolor when cut, you can chop off a piece any time and the remaining portion of the root remains crisp and watery.

Fresh jicama tastes very much like water chestnuts, but with a slight hint of sweetness. The tubers can be used in a multitude of ways. Probably the most popular use is as a substitute for water chestnuts in oriental stir-fry dishes. Sliced thinly and marinated in a soy sauce dressing, the tubers make fine appetizers, rich in vitamins A, B, and C, as well as calcium and phosphorus. They're also tasty served with your favorite dip. For a more creative appetizer or party food, squeeze fresh lime juice over jicama slices, then dip them in salt seasoned with chile powder, and garnish with pimiento, sliced red or green chile peppers, or grated cheese.

Chopped jicama is a terrific addition to a cottage cheese and fruit salad, seasoned with lime juice. The crunchy jicama provides a pleasing contrast to the soft textures of fruit and cottage cheese. Another nice way to use jicama is to add bits of it to your regular potato salad recipe. Grating the tubers into enchiladas makes for another unique dish.

Seed for the versatile jicama is available from at least three suppliers in this country: Exotica Seed Company, Horticultural Enterprises, and J.L. Hudson, Seedsman.

• The main way we use jicama is the way I originally started out with it, although we have since developed other ways, and that is as a water chestnut substitute. Even if you purchase jicama at only 49 cents per pound, it is incredibly cheaper to use than those ridiculously priced little cans of water chestnuts. The taste is almost identical, except that I think that I actually prefer the taste of jicama—it seems to be more of what I want in water chestnuts than the chestnuts themselves!

—*Cathy Bauer*
Hathaway Pines, California

Kale and Flowering Kale

Brassica oleracea var. *acephala*

THIS member of the cabbage family, also called cole or borecole, is very similar to its cousin the collard green, except that it lacks the collard's tall stem. Like collards, it is actually a primitive form of cabbage. The *acephala* part of its botanical name means "without a head." Kale is a Scottish word which came from the Greek and Roman words *coles,* and *caulis,* which they used to refer to all members of the cabbage family. Yes, kale was grown by both those ancient peoples, 2,000 years ago, and has been cultivated in more or less the same form ever since. The Romans, in fact, grew several different kinds of kale: one type had large leaves and stem, and a mild flavor; another kind was small, with small pungent leaves; some had curly leaves, and others had wide leaves like collards. The Roman philosopher Cato, in his *De Agricultura* described several varieties of kale grown by his countrymen. Kale was one of the earliest known vegetables in Europe as well. It has been grown in England since Anglo-Saxon days.

The first mention of kale in America was in the late seventeenth century, but some experts believe that since it was so popular in Europe, it actually made its appearance on this side of the Atlantic somewhat earlier.

Of all the easily grown greens in the vegetable kingdom, kale ranks today among the most overlooked and underrated

217

foods in the United States, and it's hard to understand why. Kale provides fresh greens at a time of year when other fresh vegetables are few and far between. It withstands frosts that wipe out less hardy vegetables. Nutritionally, it's at the head of its class, offering lots of vitamins A and C, as well as substantial amounts of vitamin B_2 and assorted minerals. And an average size portion contains only 50 calories, making it a boon to dieters.

Tastewise, if you're a greens lover, kale also rates high marks. It's never bitter, and cold weather actually does bring out its best flavor, just as the blurb on the seed packet promises.

Kale is certainly the easiest of greens to pick and clean. Its long, crisp, brittle green leaves are tender no matter how large they grow; even the leaf stalks cook up juicy and delicious, rather than pithy.

There are a number of varieties of kale available today. Cultural requirements are the same for all of them; the chief distinction among them seems to be leaf color. The blue-curled strains, with finely curled blue green leaves, are a bit lower growing and a few days earlier to mature. The SIBERIAN strains are taller and more spreading, with less frilled, greyish green leaves. Taste is another point of difference, but which variety tastes best will probably always remain a matter of staunchly defended personal opinion. For example, in some quarters, VATES kale is held to be sweeter, milder, and more tender than the rest, but other gardeners disagree. Similarly, to some tastebuds, Siberian kale is better used for animal fodder, while to other sets of tastebuds it tastes just fine.

Kale can be grown throughout the United States and even in Alaska. It is a cool-season crop which takes only 55 to 65 days to mature. In the South, a fall sowing provides greens in late winter and early spring. In the rest of the country, early-spring and late-summer plantings give it an opportunity to mature in the cool weather it favors.

Seed should be sown ½ inch deep, in rows 18 inches apart. The best flavor and tenderest leaves are produced by rapid growth, which can only occur in rich soil. A fertile sandy or clay loam rich in humus will yield the best results, but kale will grow

I have a particular method of harvesting kale. All summer I harvest the outer leaves by snipping them with a scissors or breaking them off by hand. If harvesting pressure is heavy, the outer leaves remain tender. When the pressure is light, the outer leaves get tough—then chickens and cows appreciate them more than people, and I harvest the next-to-outer leaves for human consumption.

When winter comes, growth slows down considerably. This makes for tough outer leaves. So once my August plot begins producing and I know I'll have a continuous supply through next spring, I engage in a superlative treat. I harvest well in any good garden soil. Kale thrives on plenty of calcium, and appreciates nitrogen as well. If you give it a good initial feeding of manure or compost, and mulch when the plants are approaching maturity, your kale will be care-free right on through late fall.

Kale is considered best after it has had a touch of frost, and least desirable in midsummer, when the leaves are less flavorful and somewhat tougher and stringier. However, harvest begins when the outer leaves are well developed but before they mature fully; if you planted a late spring crop, it means you'll begin harvesting in summer. To harvest kale, break or cut off the outer leaves as they become ready, allowing the inner leaves to keep on growing. This method will give you a continuous harvest, instead of the one-shot-and-done yield you get if you cut the entire plant at once.

Given a mild winter, kale will survive until the following spring. But where winters are so severe that an unprotected kale plant won't survive, it's easy enough to make a lean-to cold frame along the southern wall of a house, or next to any other structure that will keep the ground around the plants from freezing. Kale can take a lot of frosts, even brief dips into the teens, but as a rule it will not stand up long to a continuous hard freeze.

Even if your kale does last through the winter, you'll probably want to start a new crop in spring to be sure of a healthy new harvest. As the young shoots appear, pull up the old row and toss the plants on the compost pile or, if you have livestock or poultry—especially rabbits and chickens—they'll have a gourmet feast on those old plants.

Once harvested, kale is a good keeper. It doesn't wilt nearly as quickly as spinach or lettuce. Fully grown kale is generally considered too coarse to eat raw, but the tender, small shoots are good in salads and sandwiches. When cooked, kale has a rich flavor all its own, with a hint of the cabbage family, but its odor while cooking is much less pungent than other cabbages. It keeps its green color nicely when cooked.

Kale can be served like any other green: cook it briefly in boiling salted water, and flavor it with butter, onion, a hot

the entire growing top of plants from the first sowing, strip off the outer leaves, and save the inner rosette for a tender, mild-tasting winter salad. (By the way, if you're growing kale for the market, the entire top is the portion sold in most instances.) The tougher leaves don't go to waste. My wife uses them in hearty winter soups. Meanwhile, the second crop is growing stronger. By the first signs of spring, it's beginning to provide tender greens at a time when even the native wild vegetables are barely poking shoots above ground.

—*James Jankowiak*
Eureka, California

cream sauce, or your favorite seasoning. A traditional Irish way to serve kale is in a dish called colcannon. To make colcannon, cook a pound of finely chopped kale until tender; cook separately an equal amount of potatoes, along with a couple of carrots if you wish. Drain all the vegetables when they're done. Meanwhile, simmer some chopped leeks or green onion tops in enough milk or cream to just cover, until soft. Mash this into the potatoes, then add the kale and blend it all together. Colcannon is served warm, with a well of melted butter in the middle, and a more satisfying dish on a cold winter's night is hard to imagine.

Kale seed is available from just about any seed house.

FLOWERING KALE

The brightly colored leaves of this special kind of kale make it a favorite with many gardeners. Flowering kale is usually discussed along with its close relative, flowering cabbage, and the information presented here applies to both of these delightful plants.

Flowering kale is a long-season plant, requiring 80 to 90 days to mature. In any of the cool zones, of course, this means you have to start the plants indoors and set them out in the garden after they have a six-week to two-month start inside where it is warm. Sow as you do any of the common varieties of cabbage, in soil that is very warm, about 80°F. Using pasteurized or sterilized materials for the potting mixture will prevent damping off. Transfer the young plants into the sunlight when the green leaves appear. If you live in an area where starting indoors is necessary, you might prefer to raise flowering kale in your greenhouse and save your garden space for hardier crops. Like other members of the cabbage family, it can be grown successfully in containers under glass.

In warmer regions, flowering kale can be sown directly into the ground. Plant the seed ¼ inch deep, in rows spaced 2 feet apart. After the plants are a few inches high, thin them to stand 8 inches apart.

• At first I was leery of bringing whole heads of kale and cabbage into the house for fear of a strong cabbagey smell. It passes off very quickly, but it is important to keep changing the water and refreshing it.

—*Catharine O. Foster*
Bennington, Vermont

Flowering kale appreciates a rich garden soil, and will also respond well to some extra water if the season tends to be dry. Although subject to the same pests as other cole crops, one New England gardener reported that surprisingly, her crop was only mildly invaded by cabbageworms, and a few tomato leaves placed on top of the plants whenever butterflies were spotted in the vicinity seemed to effectively discourage the invaders.

When small, the colorful pink and white, green and white, even yellow and white heads of kale and cabbage make a surprising and delicious addition to a cabbage salad. Some growers are reluctant to use these young heads, however, because they are so delightful and decorative growing in the garden. You can also bring them in, later in the season, for floral arrangements, and very simple they are, too. All you have to do is put them in enough water to keep them alive, and their design and color are enough by themselves to make a very attractive centerpiece or decoration. Flowering kale has a slight edge over flowering cabbage in indoor arrangements, because of its curly leaves.

At the end of the season, after you've gotten all the visual pleasure out of these interesting plants, you can still slice them up for hot slaw or boiled beef and cabbage. (Cook them with some caraway seeds to reduce the cabbagey smell.)

Flowering kale and cabbage have lots of vitamins A and C, and a sizeable quantity of calcium, iron, and other minerals.

Seed sources include: DeGiorgi Co.; Glecklers Seedmen; Gurney Seed and Nursery Co.; Nichols Garden Nursery; Geo. W. Park Seed Co.; and Seedway.

Kohlrabi

Brassica caulorapa

THE origin of this odd-looking member of the Brassica family is a matter of some debate in botanical circles. Some plant historians believe kohlrabi existed in prehistoric times, others claim it was not known until the Romans cultivated it, and still others say that it was developed from marrow cabbage during the late sixteenth century.

Whatever its cultural beginnings, the name is certainly a modern one, derived from the German words *Kohl,* meaning cabbage, and *Rabi,* meaning turnip. The plant does share characteristics of these two vegetables: Its edible, above-ground stem enlargement resembles a turnip in shape and flavor; the leaves that grow from this short bulbous stem resemble those of cabbage.

The plant is easy to grow, remarkably productive, and is an all-around perfect garden vegetable. A cool-season biennial, kohlrabi demands a sunny garden bed with slightly acid to neutral soil, rich in organic matter and moisture. A light, sandy medium will encourage rapid growth and make for very tender kohlrabi. An additional application of compost increases the amount of available plant nutrients and improves the water-holding capacity of the soil.

Like cabbage, kohlrabi can be started indoors or in the cold frame and transplanted to the garden after most danger of frost has passed. But except in northern regions, the plant is usually seeded directly in the open bed.

Prepare the bed as early in spring as the ground can be worked. Loosen the soil with a spading fork, then spread finely ground limestone as needed to create a pH of preferably not less than 5.5. Next, add 2 or 3 inches of compost and turn under. Finally, spread a little garden soil over this and rake it level. Sow seed thinly, ½ inch deep in rows 10 inches wide, and 15 to 18 inches apart. Since their top growth is sparse, mature kohlrabi will not shade nearby row crops. The bulbs mature quickly, so squash planted on either side of the rows will have enough room as soon as the earlier crop is harvested. Onions and beets are also excellent companion crops, for they occupy different soil strata and can be grown in the same row as the kohlrabi.

After a few weeks, thin the seedlings to stand 6 inches apart, and mulch the bed with hay, compost, or grass clippings. Although kohlrabi is more drought-resistant than such crops as turnips and radishes, it reaches optimum succulence and tenderness only when supplied with adequate moisture. If you fail to provide sufficient fertilizer and water, your kohlrabi will probably turn out to be tough and woody.

The plant is extremely quick to mature—it can usually be harvested within 12 weeks of seeding. The popular EARLY WHITE VIENNA matures in 55 or 60 days, while the PURPLE VIENNA requires 65 to 70. When the swollen stem is 2 or 3 inches in diameter, it is ready to harvest. Larger bulbs are especially crisp, but they do not taste as sweet as the smaller ones. Successive plantings at two- to three-week intervals from early spring through June will ensure a constant supply of tender kohlrabi. An average yield is about 80 pounds of trimmed kohlrabi per hundred feet of row.

To harvest, cut the stem about an inch below the "bulb," and tie the plants in bunches like beets. Though they are edible, the leaves are usually discarded prior to cooking or storage.

An excellent source of calcium, phosphorus, iron, vitamin C, vitamin A, and several B vitamins, kohlrabi is a vigorous, easy-to-raise crop. Given plenty of water and nutrients, it is rarely bothered by insects or diseases. Cabbage loopers may put some holes in the leaves, but the worms are easy to see and pick off and they never feed on the bulbs. If the problem is especially serious, *bacillus thuringiensis* can be applied. Aphids might also attack the plants, but a hard spray of water from the garden hose will eliminate them. Companion plantings of nasturtiums and mustard will act as traps for several insects that might otherwise be attracted to the kohlrabi. Tomatoes, strawberries, and pole beans should not be planted near kohlrabi, however.

Kohlrabi fares well in underground storage along with root crops. Remove the leaves and roots, and store the bulbs in a very humid (95 percent) root cellar or basement where the temperature is about 32 to 34°F. If you lack such facilities, you can freeze your extra kohlrabi. Wash and trim the bulbs, dice them into half-inch pieces, blanch 1 to 2 minutes, cool, pack, and freeze.

Kohlrabi can be substituted for turnips in any recipe, grated raw in salads, stuffed with meat, or stir-fried in Chinese dishes. Serve it au gratin or creamed, as a side dish to accompany meats. It's especially tasty on a raw vegetable platter, served with a tangy dip for an appetizer or snack. It freezes very well and can be stored in sand in a cool, dry place for one or two months.

Kohlrabi seed can be obtained from almost any large seed supplier. W. Atlee Burpee Co.; Comstock, Ferre and Co.; William Dam Seeds; DeGiorgi Co.; Johnny's Selected Seeds; and Earl May Seed and Nursery Co. all handle both the white and purple varieties.

• I found that kohlrabi responds very well to heavy fertilization with rich compost. In our lower part of South Carolina, I plant it at the same time I plant early mustard and turnips. It grows off quickly and is very tender when planted at this time. Planted later, it matures in the hotter time of early summer and gets woody and doesn't have a good flavor.

I guess I plant it for the novelty too, but all our family likes it either raw in salads, or cooked; our son likes to have his own little garden patch so he can have a kohlrabi anytime he wants one. I tell friends who have never seen it before that it is an educated turnip, it grows above the ground. It is a regular in our garden.

—*R. L. Brewer*
Hampton, South Carolina

Leek

Allium porrum

THE leek has been cultivated for more than 4,000 years, yet this mildly flavored member of the onion family is having a difficult time establishing itself as a staple in American gardens. Other cultures, though, have held this humble food in high esteem. The ancient Egyptians regarded the leek as something of a sacred plant. Roman history has passed along the anecdote that the Emperor Nero demanded leek soup daily because he firmly believed that leeks were good for the vocal chords, and made his orations more distinct and sonorous.

The Romans are credited with distributing this native of the Mediterranean region and Near East across Europe and the British Isles, where it became so popular that the Welsh adopted the vegetable as their national plant and emblem. Tradition has it that the leek was instrumental in helping the Welshmen defeat the Saxons in a battle waged in 640 A.D. They adorned their hats with leeks from a nearby garden before going into battle in order to distinguish themselves from their enemy. To this day the Welsh wear a leek in their hats on St. David's Day to commemorate their victory.

The Englishmen in Northumberland still treat leek growing as a fiercely competitive sport, each trying to grow the largest leeks with the longest white stems. Every year, the men there become so absorbed in their leek beds that their wives

227

• Sometimes the grocer's leeks have sand down in the top leaves. The sand is difficult and annoying to remove. With care the home gardener can avoid this problem by planting the leeks in a trench which gradually becomes filled up higher during the growth of the leek. The commercial practice of hilling up around the developing leeks to blanch the stalks is the reason for the sand problem.

The failures I've had with leeks are modest. The leeks were smaller than I wanted and not well blanched. This failure occurred because I did not provide sufficient fertilizer over the long growing season of the leek. Having learned my lesson I now put down a 6-inch layer of composted manure in the trench dug for leeks. I fill the trench with a few inches of soil and plant the leek transplants on top. Infrequent feedings with manure tea will help the leeks grow large.

—John Meeker
Gilroy, California

complain of being "leek widows" the same way so many American women bemoan being "football widows" or "golf widows."

The leek's glorious history and its present widespread popularity in Europe has had regrettably little effect on American gardening habits. Many folks simply dismiss the leek as nothing more than an overgrown green onion when, in fact, it is much more. It is low in calories and high in vitamins; it repels insects; it can be harvested in the dead of winter; and as for taste, its mellow, mild flavor is beyond compare. It seems that leek and potato soup became a legend not too long after the invention of fire. One of the most tempting qualities of the leek is that it is extremely hardy and easy to grow. Many gardeners agree that, given adequate soil and care, it is just about impossible not to achieve success.

Although the leek is a member of the onion family, it does not develop a true bulb, but grows straight from the neck to the root. The white stems grow to be 2 inches thick and between 6 and 8 inches long, with large, flat grey green tops growing over a foot high.

The two most popular varieties of cultivated leeks are BROAD LONDON (also known as the LARGE AMERICAN FLAG) and ELEPHANT. Other respectable leek varieties are SWISS SPECIAL, ODIN, which is thick-stalked and of high quality, and CONQUEROR, a slender-stalked hardy strain especially noted for winter eating.

These vegetables will grow in just about any climate. In California and the southern United States, leeks are started in summer and grown throughout the winter. In the rest of the country, they're started in spring and grown through the summer. While they prefer a deep, rich, well-drained clay soil with plenty of moisture, a gardener from arid Oklahoma reported obtaining a quality leek crop just by giving them frequent waterings, a location with afternoon shade, and a good dose of rich compost. She was so impressed with their durability that she marked them down on her list of "care-free" vegetables. Even gardeners in the colder northern states should have success with leeks since they are entirely winter-hardy.

Unlike onions and garlic, leeks are generally grown from seeds. Because they have a long growing season (four to five months) they should be sown as early as possible. One very successful method of cultivation is to arrange for the leeks to grow in a shallow trench or furrow in the garden. The bottom part of the trench should be rich in well-rotted manurings; the sides are gradually filled in on the leeks as they grow to blanch them. Blanching is necessary to make the leeks more delicate and tender. If the soil is too loose, you may find that the sides of the trench erode quickly, smothering the young plants on the bottom. To resolve this problem, make a wider, flatter side to the slope of the trench.

In early spring, when the soil can be worked, deeply dig the bed or row where the leeks are to be planted, turning in as much manure as possible. To avoid burning the young seedlings be sure the manure is well-aged. Bone or fish meal that contains phosphates will also improve the condition of the leek bed.

Seed the leeks directly in the garden or in flats by barely covering the broadcasted seeds with friable soil. Keep moist but not soggy for two weeks. By then the seeds should have germinated. As soon as the seedlings start crowding each other, and while there is still some spring moisture in the soil, transplant them to the garden row, placing the seedlings in the trench about 4 inches deep and 6 to 8 inches apart. When the leeks are finger-sized, pull some soil from the sides of the trench to cover the bottom of the vegetable. Be careful, though. If this operation is overdone the leeks will smother. Normally an inch at a time should suffice, for the rains or your watering will wash some of the soil from the sides of the trench down into the bottom of the row. Once the leeks are 7 or 8 inches high, the trench should be gradually filled in around the stems up to where the leaves diverge to effect blanching. The leeks will also appreciate a dose of fish emulsion or a shot of manure tea periodically.

While the growing period to maturity is listed at between 150 and 190 days, leeks can be used before maximum growth is reached. If perchance an autumn frost passes through, leek-loving gardeners remain calm because they know that leeks defy cold weather. When the thermometer drops below freezing for an extended period of time, simply cover the entire plant with mulch. After the temperature climbs back above the teens and twenties, pull the mulch back, providing there is no snow on the ground. If properly protected, leeks may be left in the ground to winter over without any harm. The cultivated leek is generally a biennial plant. During the second year in spring, following a period of vegetative winter growth, it produces a large flowering scape, up to 3 feet in height and topped by a globe-shaped flower head. If leeks are left in the ground until they bolt to seed with the next growing season, they are virtually worthless as food. The seed stalk develops in the core of the leek stem and the outer layers provide nourishment for seed development. However, an interesting thing happens underground during this seed build-up: corms are produced at the base of the bolting leek. These corms when planted in a freshly prepared bed will grow to become next season's leeks. If the corms have grown up through the 8 inches or so to the top of the soil, they can be transplanted equally as deep, and you thus avoid the trouble of raking in soil to blanch the leeks.

The leek has no enemies. The hardy plant not only avoids all skirmishes with insects but repels the pests from other less fortunate vegetables whose names are on the insects' menus. Traditionally the leek, like its other onion relatives, has been interplanted between rows of other vegetables to guard against insect attacks. They are excellent neighbors for carrots since they ward off the pesky carrot flies. The leek is also a good crop to grow when the garden is troubled by clubroot.

The best time to dig up your leeks is when the cook is ready to use them. Of course they store excellently under refrigeration, but why clutter up the refrigerator when the ground does just as well? Some say the best method of storing leeks for a long period of time is to concoct the leek dish, such as leek and potato soup, and store the soup rather than the leeks themselves.

As for eating, leeks are slightly sweeter and possess a

milder flavor than the stronger onion or the pungent garlic.
They make a great substitute for onions in numerous dishes,
especially if you are interested in shedding a few pounds. They
are very low in calories (average about 25 calories a serving) and
have about half as many carbohydrates as onions. They also
boast of a relatively high amount of vitamins A and E (vegetables have never been very strong in the vitamin E department).
One serving of leeks can give an adult approximately one-fifth
of the daily requirement of vitamin C, which is nothing to scoff
at considering the scarcity of fresh vitamin sources in the
middle of winter.

Leeks may be eaten raw, alone or mixed with salads,

LEEK AND POTATO SOUP

6 leeks, cleaned
4 medium potatoes, thinly sliced
1 large stalk celery with leaves, chopped
1 tablespoon oil
4-5 cups chicken stock
1 cup milk or cream
2 tablespoons sour cream or yogurt
   freshly ground pepper, to taste
1 tablespoon fresh parsley, chives, or watercress chopped

Thinly slice the white part of the leeks. Saute leeks, potatoes, and celery in oil for 5 minutes. Add stock and simmer,
covered, until tender, about 20 minutes. Puree the vegetables
in a blender or food mill, and return to the pot, with the
stock.

Add the milk or cream and the sour cream or yogurt, and
heat through. Do not boil. Season with pepper and serve
garnished with parsley, chives, or watercress. This makes 6
servings of a rich, creamy, utterly delicious soup that can be
served either hot or cold.

steamed, lightly fried with other vegetables or broiled. They can also add some zest to any soup or stew and, of course, are one of the headliners in potato and leek soup. Leeks make a marvelous quiche, too. (They are mild enough to be served *a la grecque.*) Sometimes leeks are served creamed, au gratin, or otherwise sauced like asparagus, and therefore have earned the dubious distinction of being referred to as "poor man's asparagus" because they are relatively cheap.

Leek seeds may be purchased from many suppliers, including W. Atlee Burpee Co.; Comstock, Ferre and Co.; William Dam Seeds; DeGiorgi Co.; J.A. Demonchaux Co. (who carry 4 varieties); Farmer Seed and Nursery Co.; Gurney Seed and Nursery Co.; Joseph Harris Co., Inc.; Charles C. Hart Seed Co.; H.G. Hastings Co.; J.L. Hudson, Seedsman; Johnny's Selected Seeds; J.W. Jung Seed Co.; Earl May Seed and Nursery Co.; Meadowbrook Herb Garden; Mellinger's, Inc.; Nichols Garden Nursery; L.L. Olds Seed Co.; Geo. W. Park Seed Co.; R.H. Shumway Seedsman; Stokes Seeds, Inc.; and Thompson and Morgan, Inc. Le Jardin du Gourmet and Well-Sweep Herb Farm sell small transplants ready to set out in the prepared bed.

• The real pleasure in growing leeks is harvesting them on a winter day—a day when you have forgotten the bounty of September, a cold day when the garden is dormant, a day perhaps when you haven't even walked in your garden for a while. You'll need a shovel (I use the posthole type), a pitchfork, and a very sharp, long-bladed knife.

Toss back the mulch with the pitchfork. Make two or three incisions around each plant with a posthole shovel. Shovel out both the leek and its whole clod of soil. If you try to pull out the leek by itself and the ground is wet you will break the leek. Cut the soil away from the leek with your sharp knife. Trim off any bruised outer parts of the stalk. *Voilá,* you have leeks that are ready to be washed and cooked, and with a minimum of dirt in your sink.

—Aneta W. Sperber
Bloomington, Indiana

Luffa

Luffa aegyptiaca; L. acutangula

THIS unusual member of the cucumber family is seldom seen growing in America, but spa-bathers and boat scrubbers are undoubtedly familiar with the sturdy "vegetable sponges" it produces. Known also as dishcloth or towel gourd, luffa is widely cultivated in the tropics for food and industrial purposes. Small, young luffa fruits make excellent eating, as do the leaves, yellow blossoms, and seeds. Mature gourds are dried for bath sponges, kitchen and household scrubbers, or used in the manufacture of slippers, mats and baskets.

There are basically two types of luffa, the ridged or *sing-kwa* species and the smooth *sze-kwa*. The ridged form (*Luffa acutangula*) grows wild in India and Java and is cultivated as a food crop in most of the oriental countries. It produces dark green, ribbed, club-shaped gourds of about 1 foot in length, with wrinkled, winged seeds. The smooth luffa (*Luffa aegyptiaca*) yields much larger gourds that are light green in color, cylindrical, smooth-skinned, and contain smooth, wingless seeds. Listed variously as *L. cylindrica, L. macrocarpa,* and *L. marylandica,* it is the most common form of sponge gourd and the best variety for American gardeners. It is widely grown in its native Asia, Africa, Australia, and Polynesia, as well as in South and Central America.

As a tropical plant, luffa requires a long, preferably hot, *235* growing season. In some southern regions of this country it can be direct-seeded and easily harvested before the first frost. Elsewhere, seeds should be started indoors and transplanted to the garden bed when the soil has warmed. To direct-seed, plant seeds ½ inch deep in hills 6 feet apart, two seeds per hill. Like other cucurbits, luffa responds well to trellising; many growers prefer to plant seeds singly, at 1-foot intervals along a trellis or fence. Yields of supported plants are double, or even triple, those of rambling ones.

In cooler regions, start seeds indoors several weeks before the last frost. Sow seeds in peat pots, two seeds per pot, in a loose, peat-rich soil mixture. Again, they should be planted ½ inch deep. Because they are very hard-coated, luffa seed may require two or three weeks to sprout. Extra heat speeds germination as does an overnight soaking of the seeds. Keep newly planted pots in a protected area until the seeds sprout, then move them into the light and water well. When plants have grown to about 2 inches, thin to the one sturdiest plant per pot.

Seedlings can be set outdoors after danger of frost has passed and the plants are 3 or more inches high. For best results, water thoroughly the night before and plant on a cloudy day or in the evening. Like cucumbers, luffas do best in a light soil that is well-supplied with humus. Avoid over-fertilization, for it may lead to lush green plants with very few gourds. It's better to supply moderate amounts of nitrogen through applications of well-rotted manure or compost. Full sunlight is necessary and the soil should not be allowed to dry out.

For optimum yields, begin training the vines immediately, using strings where necessary to fasten the vines to the fence or trellis. If you do allow the plants to ramble over the ground, protect the developing fruits from the damp soil with a board or clean, dry straw.

The vines are vigorous growers with large, green leaves and deep yellow blossoms. Leaves can be harvested at any time to be used in salads or cooked as greens, and the large flowers can be dipped in batter and sauteed. In midsummer the dark green fruits begin to appear. If the plants' first flowers have been removed, these subsequent fruits will be well-formed and of fairly uniform size. Young fruits can be picked when under 6 or 7 inches long, before they become stringy, and cooked like okra or summer squash. They can also be sliced and used in salads instead of cucumbers. Left to mature for sponges, they attain lengths of 15 to 24 inches and weigh several pounds. Any poorly shaped or rotting fruits should be removed as they appear.

For high-quality vegetable sponges, allow the luffas, or loofahs as they are sometimes called, to ripen on the vines. They are ready to harvest when the stem yellows and the skin begins to dry and fade. The greener the skin, the more tender the resulting sponge; very yellow luffas may make for too-wiry sponges.

Even if seeds are started indoors, many northern growers find their luffas do not have time to reach this mature stage before the first frost. In these zones, plants must be raised in the greenhouse. Sow seeds directly in the greenhouse bench or in peat pots for transplanting. Trained on wires fastened to sash bars, the plants make productive use of waste space and are very lush and beautiful foliage screens. The heat and humidity of a greenhouse environment are, of course, perfect for this tropical plant, and the fruits develop quickly, reaching maturity within several months.

To harvest, cut the gourds from the vine with a sharp knife. Dry them for about two weeks, until the skin hardens and turns brown, then open the large end and shake out the seeds. Soak the sponges overnight, then peel off the outer skin and allow them to stand in the sun to dry.

For softer sponges, place harvested gourds in a kettle of water and boil for several minutes. "Zip" off the outer skin by pulling on the fibrous "strings," then wash out the centers, remove all loose tissue and seeds, and dry gradually in a shady place. For white sponges, soak dried luffas for one-half hour in a weak solution of household bleach.

The luffa sponges are highly esteemed by skin specialists who recommend them for the removal of rough skin and to promote blood circulation. In the kitchen they are used to scrub

• To prepare stuffed luffas, cut lengthwise, scoop out a cavity and fill with a mixture of cooked rice, onions, carrots, sesame seeds, thyme, and sweet marjoram which has been lightly sauteed. Bake for 30 to 40 minutes at 350°F.

Luffa soup provides a hearty lunchtime dish. Cut one 6- or 7-inch gourd into chunks or slices (do not peel or remove seeds). Place in a saucepan and add 2 cups water and a small onion (chopped). Bring to a boil, cover and simmer 15 minutes. Pour hot mixture into blender, add 4 fresh basil leaves or 1 teaspoon dried basil, and liquefy. Season to taste, and serve with slices of whole grain bread and a garden salad.

—Bonnie Fisher
Peterstown, West Virginia

vegetables, pots and pans, and as dishcloths. Around the house they come in handy to scrub down painted surfaces. Indeed, this versatile vegetable earns its popular name of "Nature's Scrubber."

Luffa seed is available from Exotica Seed Co.; Gleckler's Seedmen; Grace's Gardens; Gurney Seed and Nursery Co.; J.L. Hudson, Seedsman; Johnny's Selected Seeds; Lakeland Nurseries; Nichols Garden Nursery; Redwood City Seed Co.; R.H. Shumway Seedsman; Sunrise Enterprises; and Thompson and Morgan, Inc.

• Last year I devoured my first loofah gingerly, but later ones with real relish. My favorite was to parboil the gourd cut up, skin and all, then to finish cooking it *en casserole* with tomatoes, topping it with grated Parmesan cheese. It looks rather like zucchini, is not as tender. This year more cooking experiments will be made.

—*Clara Cassidy*
Harper's Ferry, West Virginia

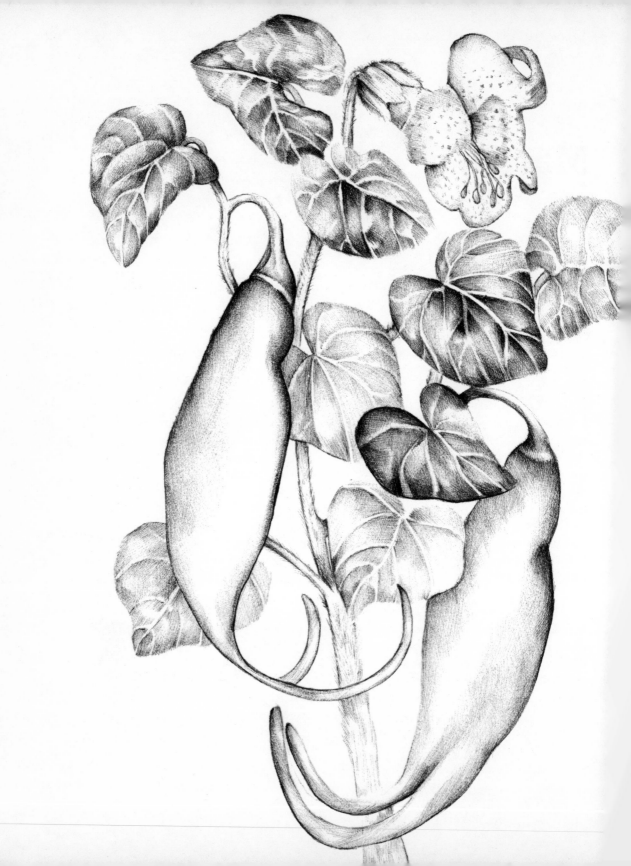

Martynia

Proboscidea louisianica, P. spp.

NINETEENTH-CENTURY American gardeners grew martynia both for its showy flowers and its unusual, edible seedpods. Native to southwestern North America, it is a warm weather annual with heavy stems, coarse roots, and soft, crinkly foliage. Leaves range in size from 6 to 12 inches and are bristled so that water rolls easily off the surface. Flowers come in various shades of purple, pink, and yellow with contrasting speckles on the upper "lips." At the end of every stem as many as 12 fragrant blossoms may form, each one measuring at least 2 inches across. The edible seedpods that follow resemble curved okra pods. They are soft and green when young and can be pickled or used in soups. Later, they ripen into extremely hard, brown capsules with sharply curved horns. These curious pods have given the plant many fanciful names, like Unicorn Flower, Proboscis Plant, Devil's Claw, and Ram's Horn.

Martynia belongs to the family Martynacia; the genus is listed as *Proboscidea* or *Martynia.* Four species are found within the United States and several others, although native to Mexico or Brazil, can be grown here. *P. fragrans,* a crimson-flowered species, is fairly common in Mexico and is a favorite for pickling; *P. lutea* is an old-fashioned Brazilian variety with yellow flowers. Of the American species, three are most commonly found: *P. annua* sports copper-colored flowers; *P. parvi-*

241

flora comes in yellow or purple; and *P. louisianica* (also classified as *P. jussieus*) has pale pink, yellow, and orchid blossoms. All of these will grow in almost any zone, but northern gardeners have had the most success with *P. louisianica*. Native from Delaware eastward to Indiana and south to New Mexico, it is the one offered by most of the eastern seed suppliers. *P. fragrans* is often grown under this name.

Martynia requires about four months of warm weather to achieve full maturity. Southern gardeners can sow seeds outdoors in the spring, but elsewhere plants must be started indoors or under glass. To direct-seed, sow 1 inch deep in rows 3 feet apart, later thinning the plants to stand 18 to 24 inches apart within the row. Transplants should be started indoors in late February or March. Sow the seeds in flats of sandy sterile soil or soil mix and cover them with about 1 inch of the medium. Water thoroughly and place in a warm, humid location until germination occurs. When the first pair of leaves has formed, transplant the seedlings to individual pots. Keep them moist and slightly shaded until the roots are well established. Martynia plants can be set outdoors, along with tomatoes, eggplants, and melons, as soon as the soil has warmed and the danger of frost has passed. Space them at 4-foot intervals in rows 3 to 5 feet apart. Soil should be light, but moist, and very rich. Full sunlight is preferred.

The plant grows to a height of 2 feet and is rather wide-spreading. Its hairy stems tend to wander off in all directions, but if controlled, it makes a particularly striking border. Martynia needs no support or special care other than weeding, fairly frequent waterings, and a good mulch of compost or decayed manure. If proper care is given, flowers will appear about six weeks after the seed germinates. They drop from the vine unwilted and cover the garden bed with a carpet of bright petals. Seedpods follow immediately, forming behind the blossoms and hanging in clusters from the heavy stems. The pods can be harvested for eating when the plant is about two months old, when they are still green and fuzzy. If gathered while tender they make good vinegar pickles. Whole or sliced, they are intriguing additions to the soup pot.

Seedpods left to dry on the vine gradually harden and split until two dark horns form at the beaked end. The outer skin drops off and the black seeds fall, leaving a "devil's claw." Wired to curving stems or hung by their own natural hooks, these pods add striking touches to dried or fresh flower arrangements and wreaths.

DeGiorgi Co.; J.L. Hudson, Seedsman; Geo. W. Park Co.; and R.H. Shumway Seedsman carry martynia seed.

Mustard and Mustard Spinach

Brassica juncea

THE beautiful, bright flowers of the mustard have long been a familiar sight to country dwellers and travellers. In meadows and fields, along roadsides and railroad tracks, seas of yellow flowers splash the springtime landscape with their brightness. The many-flowered wild mustard and its cultivated relatives are all extremely useful plants. Mustard fanciers are well aware of the plant's worth as a vitamin-rich salad green and potherb, a therapeutic medicine (the source of the old-fashioned mustard plaster), and, of course, the most popular condiment this side of catsup. Mustard has been cultivated throughout the temperate world for the last 2,000 years or more, and is oft-mentioned in the Bible, and in the literature of the ancient Greeks and Romans. Unfortunately, mustard is too often neglected by today's American gardeners in favor of spinach.

The mustard which is grown in the United States for greens, and with which we'll concern ourselves here (*Brassica juncea*), should more properly be called "India mustard." Mustard developed in many parts of ancient Asia, and this type is descended from strains that are believed to have evolved on the Indian subcontinent. They are the large-leaved, pungent mustard greens which are used in salads and as potherbs. Two closely related varieties, white mustard (*B. alba,* or *Sinapis alba*),

245

and black mustard (*B. nigra* or *S. nigra*) are usually grown for their seeds, though the young leaves can be eaten as well. Of the two, white mustard is considered to be mellower in flavor, and is often found as half of the traditional British salad called "mustard and cress."

India mustard has evolved into many different varieties, several of which are grown in this country and will be discussed shortly. There are also a number of oriental mustards, the best-known being Bok Choy, which are similar to India mustard in the quality of their leaves and in their growing habits. But these Far Eastern types are a different species, *B. japonica,* and are discussed in this book in the section on Asian brassicas.

India mustard, which from now on we'll call simply "mustard," is a hardy annual of the Crucifer family. Like all mustards, it develops spikes of four-petalled yellow flowers when left to bloom. Closely following the flowers are small, spear-shaped seedpods approximately ½ inch long. Mustard which is left to grow to maturity and reseed itself for a few generations will revert to the state of its strong-growing and strong-tasting wild predecessor. Depending on the species, the plant can attain a height of anywhere from 2 to 6 feet. There are several varieties of mustard available from American seed houses, varying in both appearance and degree of pungency.

Perhaps the best-known type is SOUTHERN GIANT CURLED, which generally grows to about 3 feet tall, and has broad leaves with frilly edges. This is one of the most pungent mustards. A newer strain listed as an improved version of Southern is GREEN WAVE, whose bright green leaves also have a hot flavor. This strain is said to do well in spring plantings and to resist bolting. PRIZEWINNER is another pungent, long-standing curled mustard.

For sheer beauty, the mellow-flavored OSTRICH PLUME or FORDHOOK FANCY wins hands, or rather, leaves, down. Its tightly curled, feathery leaves are favored for adding an attractive touch to salads, and have a pleasantly mild flavor when cooked.

A favorite of many gardeners is FLORIDA BROAD-LEAF, sometimes called MUSTARD CABBAGE, which sports broad, succulent leaves with a large, tender rib. The smoothness of the leaves makes this variety easier to clean than its curly counterparts. Florida Broad-Leaf has been known to grow to a rather remarkable size without becoming tough or too hot. One Oklahoma gardener boasted a plant with a 4-foot spread and leaves nearly 2 feet long and more than a foot across. She declared that every inch was tender and delicious when cooked.

Yet another popular variety of mustard is TENDER-GREEN or MUSTARD SPINACH. This type is treated as a separate vegetable in many seed catalogs, and will be discussed separately at the end of this section.

Some of the mustard grown commercially is actually an imposter called rape. Rape (*Brassica napus*) has seedlings whose leaves are a more intense shade of green than those of India mustard. Its seeds are cheaper and it stands up to hot weather better than mustard, although it lacks the pungent flavor.

Mustard is a hardy, cool-season, short-day crop, so proper planning is essential to its success. As early as mid-April you are passing out of prime planting time which won't come again until August or September for the autumn crop. So plant early. Even in northern zones, it's safe to plant seeds two to three weeks before the last expected frost. If you live in a milder climate further south, planting can take place all winter long. The pungency of mustard leaves increases with the daily temperature. If planted when the weather is too hot and the days too long, mustard will inevitably get too hot and peppery to eat by harvest time. More than likely the plant will also bolt to seed before you get a chance to enjoy it.

Mustard isn't too particular about its soil. It prefers a sandy or any other light loam, but will prosper in almost any condition. Average soil and average fertility will do fine. Mustard will even tolerate mildly alkaline soils. Save the bulk of your compost and manure for the main-season crops that will need it after the mustard goes out. About the only thing it will in no way endure is dryness. If you water regularly you'll get mild, crisp greens, but if you are inattentive and forget, one bite will send you sprinting for the water faucet.

The seeds should be sown thickly a half inch deep in drills

• Along the Pacific Coast, I plant mustard outdoors in the beginning of February. If the fog cover remains, I'm able to plant successive crops up to June. Come late August, I start again—and the crop grows until the rain sets it back sometime in December. During mild winters the mustard keeps growing all winter, just as it does in southern California and the Deep South.

When I lived in a section of California where summer temperatures reached into the 90s and 100s almost daily, I stopped planting by early May and didn't resume until September. Whenever the weather is hot and the days long, mustard gets too hot to eat by the time it's harvestable . . . It's this summer planting that has made many people think mustard is a difficult-to-grow green that *always* has a peppery taste.

—*James Jankowiak*
Eureka, California

• Mustard is another vegetable which is ideal for natural farming. It can be left to grow untended among the weeds. The first time I saw mustard growing in this semi-wild manner it was covering the slopes of a citrus orchard. The sea of yellow flowers beneath the trees laden with ripening grapefruits and tangerine blossoms was a truly unforgettable sight. To grow mustard this way the seeds should be scattered among the weeds very early in spring before a rain which is likely to last for several days. Then cut the weeds and scatter the cuttings over the seeds as mulch to hide from birds and chickens.

Depending on the climate, the soil condition, and the composition of the natural ground cover, the weeds may or may not have to be cut back again before the mustard establishes itself. Once it does, however, it will come up and reseed itself year after year.

—*Larry Korn*
San Francisco, California

12 to 15 inches apart. Water with a fine spray and wait for germination. It is usually beneficial to water with manure tea or fish emulsion to get the mustard off to a good start. A side-dressing of blood meal adds nutrients and repels marauding rabbits. As the rows become crowded, thin the plants to stand 6 inches apart, but don't be wasteful; the thinnings are great to perk up springtime salads. Mustard is at its most tender and delicious stage when its leaves are 4 to 6 inches long. To be sure of having the best-quality leaves for the table, periodically go down the rows and snatch all leaves over 4 inches long, leaving the inner core to grow and harvest at a later date. There is a thin line between when mustard is crisp and tasty and when it is about to bolt and get too hot, so many gardeners recommend successive sowings every two weeks in early spring to keep a fresh crop coming in. For the best-tasting greens, never let mustard mature fully—the older it gets, the tougher and hotter it becomes.

Mustard is seldom bothered by pests or disease, and is easily grown year-round in the greenhouse if a temperature of 50 to 60°F. can be maintained. The seed should be sown on the surface of the soil, in firm, level beds. Finely spray the seeds, then cover them with steam-sterilized wet sackcloth. When the seedlings are 1 to 1½ inches tall remove the cloth. The yellowish leaves will turn green within a few days and like mustard grown outside, they should be cut when about 4 inches long.

This hardy brassica also performs well in a cold frame, providing greens even in severe winter weather.

The plant can also be an easy-to-care-for houseplant. Simply plant some seeds in a flower pot, place in a warm window, water amply and pick leaves when needed. Just the thing to put some glamour in an otherwise humdrum winter salad.

Whether you eat them in winter or summer, one cup of cooked mustard greens could be just what the doctor ordered: a whopping 8,120 international units of vitamin A, only 35 calories, 193 milligrams of calcium, 68 milligrams of vitamin C (more than your average orange), and substantial amounts of iron, phosphorus, and B vitamins. As evidence of its medicinal value, for thousands of years the crushed mustard seed has been mixed into a paste for mustard plasters and spread over sore muscles or congested chests.

You can grow mustard as a leaf lettuce, using the young, crisp leaves for salads, and cooking the more mature leaves the same as you would chard or spinach. Fresh mustard and parsley salad is a favorite. Young seedlings can be added to sandwiches or used as a garnish. In the South, mustard is traditionally cooked like collards and other greens—boiled with salt pork, the "pot likker" saved to be sopped up with corn bread. It can also be made into a clear soup along with some diced ham, or turned into cream soup. A nineteenth-century British recipe for "herb pie" calls for a mixture of lettuce, mustard, beet greens, spinach, parsley, borage, and watercress to be covered with batter and baked in a pie crust. One unusual way to prepare mustard is to tempura it. Dip the leaves and flowers in batter by themselves and deep fry. Or you can mix them with other spring vegetables such as mugwort, bracken fern shoots, the young leaves of the persimmon tree, or whatever is at hand. Cooking tones down the flavor, so you may find that leaves which are much too hot to eat raw will be acceptable when cooked.

A 10-foot row of mustard should produce enough greens for a family of four. If you find that you planted too much, the chickens will be more than willing to eat what you can't. The plant's abundance of vitamins and calcium means stronger eggshells, and therefore more and better-tasting eggs for you, in addition to a choice manure.

If the crop gets past you and the poultry and goes to seed, you can always make some prepared mustard to use as a condiment or medication. Collect the pods when they're just dry, before they have a chance to shatter. The seeds can be removed either by flaying and winnowing, or by opening the pods, one at a time, over a paper bag. Grind the seeds in a flour mill (a blender might work, too), add some vinegar and water, salt, and spices and you've concocted a semblance of the genuine article.

Or you can save the seeds for the next crop. One of gardening's great delights is sowing seeds collected from your own planting. As you sow the best of the best from year to year, you'll develop an improved local strain of mustard.

Mustard seed can be purchased from many companies, including Burgess Seed and Plant Co.; W. Atlee Burpee Co.; D.V. Burrell Seed Growers Co.; Comstock, Ferre and Co.; DeGiorgi Co.; Farmer Seed and Nursery Co.; Gurney Seed and Nursery Co.; Joseph Harris Co., Inc.; Charles C. Hart Seed Co.; H.G. Hastings Co.; J.L. Hudson, Seedsman; Jackson and Perkins Co.; Le Jardin du Gourmet; J.W. Jung Seed Co.; Earl May Seed and Nursery Co.; Mellinger's, Inc.; L.L. Olds Seed Co.; Geo. W. Park Seed Co.; Redwood City Seed Co.; R.H. Shumway Seedsman; and Otis S. Twilley Seed Co.

MUSTARD SPINACH

Mustard spinach, or Tendergreen, owes its existence as a separate vegetable primarily to the practice of some seed houses of taking a variety of a common species and, by putting it into its own classification, making it something more than it really is. Nevertheless, this variety of mustard does deserve special mention. It grows rapidly (maturing in about 25 days), withstands cold weather as well as BLOOMSDALE spinach, stands in the rows long after its relatives have bolted to seed, and never develops the unbearable hotness that overtakes other varieties with age. When other mustards leave you gasping for water, Tendergreen remains crisp and mild.

Mustard spinach probably got its name from the slight resemblance its leaves bear to smooth-leafed spinach. But its savory flavor is nothing like the taste of spinach. This variety is tremendously popular, and many gardeners grow it as their chief mustard crop. If you decide to succession plant for a continuous crop, you may find it cheaper to buy seed by the ounce instead of the packet. An ounce of seed, carefully doled out, will cover up to 200 feet of row—more than an average family of four will need for an entire gardening season.

In the summer, mustard spinach will keep on growing after other mustard has bolted to seed or is too hot to eat. It will also outlast long-standing spinach. This makes it ideal for families who only need a limited amount, and who also don't want to spend a lot of time making succession plantings. The outer leaves can be picked a few at a time as needed for a long time before the plants bolt. In all areas of the country except where summers are cool and moist, it is best to treat Tendergreen as a spring and fall crop, stopping plantings in early May and beginning again in late August or early September. At any time of the year except in early spring a thick mulch is sound practice. Check the moisture content of the soil regularly, too. Water the plants when the first inch of soil, or better still, the first half-inch, has dried out.

One California gardener conducted a side-by-side unheated greenhouse experiment one winter, and concluded that mustard spinach is a better choice than the other varieties of mustard for winter growing, because of its ability to stand for a long time without getting hot. In growing any crop in cold weather, one of the biggest losses in time is waiting for the seeds to germinate and establish themselves. To avoid that loss, he decided not to make succession plantings. Instead, he sowed the whole crop, and when the plants began to crowd one another, ate the thinnings. After thinning, the remaining plants stood 3 inches apart. From then on he fed the plants with seaweed emulsion at the soil surface, and for greens, the outer leaves were picked as needed. "With this method," he explains, "I've harvested mustard spinach for months in winter and early spring from the same plants with no problems with hot, pungent tastes. Southern Curled mustard, on the other hand, grew bitter with the same treatment."

Seed for mustard spinach is available from W. Atlee Burpee Co.; Comstock, Ferre and Co.; Gurney Seed and Nursery Co.; H.G. Hastings Co.; Kitazawa Seed Co.; Earl May Seed and Nursery Co.; Nichols Garden Nursery; L.L. Olds Seed Co.; Geo. W. Park Seed Co.; Redwood City Seed Co.; and R.H. Shumway Seedsman.

Nasturtium

Tropaeolum majus, T. minus

THE nasturtium is a native of Peru, brought to this country in the mid-eighteenth century. Sometimes called Indian cress and creeping canary, it became immediately popular as an ornamental and vegetable plant. Early American gardeners used the flowers in bouquets, the flowers and leaves in salads, and the pickled seeds as a substitute for capers.

There are two varieties of nasturtium used as vegetables, *majus* and *minus*. *Majus* will climb fences and trellises, often reaching a height of 8 feet or more. *Minus* is much lower growing, produces smaller fruits and flowers, and is an excellent border plant. Both can be grown in the greenhouse, the tall type along trellises, the dwarf variety in pots or hanging baskets. For outdoor garden beds, they can be started in flats and transplanted outside as soon as danger of frost has passed. Nasturtiums are semi-hardy and fast growing, however, and will perform very well if simply seeded directly in the garden in early spring. A light soil with full exposure to the sun is best, but the plants will grow in almost any garden corner. They thrive in poor soil; in fact, soil overly rich in nitrogen causes nasturtiums to develop a rank growth of foliage at the expense of the flowers.

To plant, sow seeds ½ inch deep in rows 4 feet apart. Nasturtiums do not require much water, but like most plants, they benefit from a mulch applied after the seedlings have become established.

Every portion of the plant is edible. The red, orange, yellow, or pink flowers that appear in early summer and continue throughout the season can be picked for use in salads or sandwich spreads. Their flavor is rather hot and peppery, resembling that of radishes. Nasturtium vinegar can be made by packing fully opened flowers into a bottle and covering them with vinegar to which garlic, shallots, salt, and red peppers have been added.

The leaves are quite edible and have a taste a bit like watercress. Choose the tender, young leaves at the growing tips of stems for use in salads and sandwiches. A small amount of this tangy salad herb will add zest to many fresh vegetable dishes. They can be stuffed the way Greek cooks stuff grape leaves, or added to a green salad dressed with a tangy sour cream dressing.

The seeds of the nasturtium grow as large as peas and are delicious when pickled. To make mock-capers, the seeds are gathered while green, washed well, and packed into clean bottles or small jars. A mixture of vinegar, sugar, salt, pepper, and hot pepper is boiled, then allowed to cool. The cooled solution is poured over the seeds. Jars are then sealed, and should age for six weeks before the seeds are used. The "capers" will keep for several months.

With a little foresight, fresh nasturtium blossoms and leaves can be enjoyed long after the garden has died down and the ground has frozen. Before autumn cold arrives, root some nasturtium tip cuttings in sand or perlite. When the roots have become established, repot and place the plants on a sunny windowsill where they will provide colorful and edible leaves and flowers all winter long.

Nasturtium seeds can be purchased from practically any seedsman and are even available in most storefront seed packet displays.

• Salads are a big item in our home. Dandelion, miner's lettuce, pigweed, jicama, mache, *girasole*, goosefoot, bean sprouts are just a few of the plants we mix with the usual lettuce in our salads. One of our favorites is nasturtium. The best feature of nasturtium is that it is an attractive plant in the garden, a plant bearing red, yellow and orange flowers through the summer. It has a reputation for repelling some harmful insects from the vegetable garden. Add to this the fact that nasturtium does not have to be pampered, it can reseed itself from year to year, it is tolerant of poor soils, and it does not require a lot of watering.

—*John Meeker*
Gilroy, California

Nettle

Urtica dioica

THE unfortunate nettle is among the least appreciated of plants. Many innocent people have at some time or another accidentally brushed up against its bristling hairs and felt its sharp sting, which has been likened to hundreds of hypodermic needles puncturing the skin, or a horde of ants biting all at once. Because of such unhappy experiences, the nettle has been judged as something to be avoided at all cost. But despite its rather unsociable first impression, the common stinging nettle is an extremely useful and beneficial plant to have around. Even the sting itself is a temporary remedy for rheumatism and failing muscular strength. The nettle is remarkably versatile. It can be used as a spinach substitute, for making clothes, relieving sore throats, feeding livestock, enriching compost, even promoting hair growth. And these are only a few of its functions.

The nettle has been growing in most areas of the world for centuries. In the British Isles, the plant has been adopted as somewhat of a standard ingredient in a wide assortment of dishes. A famous nettle pudding (made of nettles combined with leeks, broccoli or cabbage, and rice and then boiled in a muslin bag) comes from Scotland; a nettle cream soup is a specialty in Ireland; and nettle beer is made in some districts of Britain. There are at least seven species of nettle with stinging properties in the United States, the two most common being the stinging nettle, also called the common nettle or great stinging nettle, and the slender nettle. The common nettle apparently did not make its debut in America until after European colonization.

The perennial stinging nettle grows to 7 or 8 feet in height, has bristly hairs on its square stems and sawtoothed leaves, and small clusters of dull greenish flowers sprouting near the joints of the leaves and stems. At the slightest touch, each bristle's globular tip is knocked off, leaving a sharp point which can easily penetrate the skin. Simultaneously, a noxious liquid derived from formic acid oozes out of the hair and into the skin of the intruder. Small red welts appear shortly after the moment of contact.

But if handled with care, nettles can make a valuable addition indeed to your garden. The first step is getting them there. Nettles can be grown from seed, but it's also a fairly easy matter to dig up clumps of dormant or young sprouting plants from the wild and place them in similar conditions in the garden. Sometimes it takes a while for them to become established, but in time they almost always succeed.

Young nettle leaves make a tasty potherb, which you can prepare exactly as you would spinach. One gardener's favorite recipe is to serve steamed nettles mixed with chopped hard-boiled eggs and topped with melted butter. They can also be wilted like lettuce and served with a hot bacon dressing (see the dandelion section for a recipe), or made into croquettes. Chop and cook the nettles, drain, and combine with some chopped onion, beaten egg, melted butter, grated cheese, and bread crumbs. Roll the mixture into balls and brown them in oil. To make the traditional Irish nettle soup, cook 6 chopped leeks in butter, then add a quart of milk and simmer until the leeks are tender. Stir often, and don't let it boil. Add salt to taste, 4 cups of chopped nettles, and 2 tablespoons of cooked oatmeal to thicken the soup. Cook until the nettles are tender and serve piping hot. This recipe will serve 6.

Armed with a pair of gloves, you can begin to harvest young nettle tops when the plants are approximately 6 inches high. The

If you inadvertently brush up against a nettle plant and are stung, take heart. Here's an old herbal remedy that brings relief. It's made from the mullein, which grows even more commonly than the nettle in many parts of the country. Crush some mullein leaves and apply them to the welts for a few minutes. The pain should quickly subside.

first harvest may be as early as March, and may continue until as late as mid-June. Be sure to harvest only the growing tips, which include the rosette of young forming leaves and the terminal bud plus the two or three pairs of leaves just below the rosette. The leaves, when eaten, are tender with a rather salty, earthy flavor. It is possible to use the growing buds of older plants, but they are tougher and not as finely flavored. For those of you who fear getting nipped in the throat at the first bite, be comforted to know the nettle loses its sting once it is boiled or dried. After the harvest, nettles can be stored by drying and then placing them in airtight containers.

Throughout history the nettle has proven itself time and again to be a valuable culinary and medicinal herb. On an average, nettles contain 25 to 28 percent crude protein (dry weight) and about 32 percent fiber. They are especially high in iron and calcium, and contain more chlorophyll than almost any other plant in existence. This, combined with the fact that it possesses generous amounts of vitamins A and C, makes the nettle a natural for healing numerous ailments. Nettle tea is said to be good for kidney disorders and to improve the function of the liver, gall bladder and intestines. Owing to its abundance of iron and chlorophyll the nettle has an especially good effect on blood-building. It also works as an excellent hemostatic agent. Pressing the boiled leaves against a wound will stop the bleeding and at the same time purify the blood. Nettle tea is noted for its ability to relieve sore throats when used as a gargle, and is made into a solution used for hair tonics and controlling dandruff. There's also a type of rennet that can be made from 3 pints of a strong infusion of nettles to a quart of salt.

What's good for man is often good for other animals. In fact, nettles are the preferred food of goats, and when made into hay are excellent feed for cattle and horses. Chickens become ecstatic when offered fresh nettle and one California gardener, James Jankowiak, is convinced he has the answer to better-tasting eggs. "I firmly believe that poultry fed on a dietary supplement of fresh nettles and other medicinal herbs like dandelion, chickweed, and comfrey will be plumper, healthier, and lay more eggs of better quality than those on commercial (or

• In Austria, young nettle sprouts are esteemed as a hearty and cheap spring potherb. From an Austrian gardener comes this recipe:

Steam 2 pounds of nettles with very little water, drain off but reserve the water, and then pass the nettles through a sieve. Heat 2 tablespoons of butter, add 3 tablespoons of flour, stir well until the mixture becomes golden brown. Then add the reserved water and stir well again. Finally add the nettles, plus 3 to 5 tablespoons of sour cream. Serve with peeled potatoes or with cooked rice and a fried egg per person.

—Helmut Leithner
Gmunden, Austria

even organically derived) mash alone," he said. This theory isn't new; in Europe farmers have been incorporating boiled nettles into their mash as a chicken fattener for years.

The Europeans have also been accustomed to using mature nettle plants to make a fiber for cloth that many claim is more durable than linen. Here in America, Indians used a native species (*U. gracilis*) for the same purpose and during World War I, when cotton was in short supply, nettles were used in large quantities for making textiles. This should come as no surprise to those who know the nettle's lineage, since it is closely related to the hemp, an extremely respectable fiber source from which top-quality ropes are made. Nettle roots are reported to be used in the making of a yellow dye.

The nettle is loaded with nitrogen and therefore can make a superb "hot" vegetable material for the compost heap, when manure is in short supply. After the leaves and stems have been cut, they rot down to make a blackish brown humus that will stimulate compost fermentation. It does equally well as a general companion plant to other vegetables and herbs, stimulating their growth, enhancing their flavor, and increasing their resistance to disease.

The nettle is more than just an irritating weed; it is a benefactor to man, animals and plants. Its versatility is awesome. You can sit down and have a dish of pureed nettles, pour yourself a cup of wholesome nettle tea (being careful not to spill any on your nettle tablecloth), go out and feed your livestock with nettle hay, before going to bed gargle and wash your hair with a nettle preparation, and finally dream peacefully knowing that your garden is prospering from nettle residues returning to the soil.

If you choose to grow nettles from seed, you can order it from Casa Yerba; William Dam Seeds; Meadowbrook Herb Garden; and Redwood City Seed Co.

To make nettle hair tonic, steep the fresh leaves in a half and half mixture of water and vinegar for 30 minutes. Strain and apply the liquid as a next-to-final rinse after shampooing your hair. This mixture is supposed to help control dandruff.

Okra

Hibiscus esculentus

OKRA is one of the many cultivated plants that originated on a fertile north African plateau in the mists of prehistory. Still found wild near the upper Nile and in Ethiopia, this edible relative of the ornamental hibiscus was apparently taken across the Red Sea to Arabia and later introduced to Egypt by Moslem conquerors. As a sturdy and prolific annual, okra was carried by the Moors into southern Europe. It also traveled east to India, where it is now well-known. Thanks to the French, okra made its North American debut around the Mississippi delta in the seventeenth century, and to this day just about every home garden in Louisiana contains this staple of Creole cookery.

Also known as lady finger and as gumbo—a word taken from the Portuguese version of the plant's African name—okra is grown commercially for soup manufacturers in several southern states. It is only now becoming known as a home gardening possibility in more temperate parts of the country, where it can be cultivated successfully just about everywhere that tomatoes, melons, and cucumbers are grown.

Growing anywhere from 4 to 10 feet high, this decorative plant boasts luscious-looking pale yellow blossoms with a splash of dark red in the center. The flowers unfurl rapidly after sunrise on very warm days, yielding pollen much loved by bees. By the end of the same day, the observant gardener often can spot a tiny

263

okra pod beginning to form under a wilting blossom. Beaked and many-seeded, these creamy white to dark green capsules may be spiny or smooth. They reach a length of 4 to 12 inches at maturity, turning brown as the seeds ripen.

Okra needs about two months from seed to start yielding the tender 2- to 3-inch-long pods that make the best eating, and about four months to generate ripe, high protein seeds. Though this vegetable has cropped well as far north as Massachusetts, if your area has a short growing season and cool nights, you'll do best with one of the fast-maturing dwarf varieties. Two of these types reaching 3 to 4 feet are PERKINS SPINELESS and DWARF GREEN LONG POD, which have green pods about 7 inches long. Some taller green pods are LOUISIANA GREEN VELVET and PERKINS MAMMOTH LONG POD, while the moderately tall WHITE VELVET (or LADY FINGER) offers round, creamy-white pods that are slightly shorter. A favorite of many gardeners is the All-American winner CLEMSON SPINELESS, which will give you green 7- to 9-inch pods on medium-sized plants of 4 to 5 feet.

Warm soil is a must for okra seed, which is hard to germinate and can rot in cold, wet ground. To get your plants off to a fast start, soak the seeds in water overnight before planting. As another aid to germination, you might lightly mulch just-seeded rows with hay or grass clippings. Sow seed ½ to 1 inch deep in rows 4 to 6 feet apart for larger varieties and 2 to 4 feet apart for the dwarfs. When the seedlings are 3 to 4 inches tall, thin them to 12 inches apart (for dwarfs) and 18 inches apart (for bigger types).

In addition to well-warmed soil, okra needs an air temperature of over 60°F., so to get a jump on things if your spring is cool, you might want to start some seeds inside six weeks before the last expected frost. Transplant in late May or early June, being careful to keep a sodball around those well-developed roots.

Gumbo demands full sun but is fairly tolerant of different soils. It does best, though, in well-drained soil that's high in humus and has a pH of between 6 and 8. Mix in some compost if you have it, but go easy on nitrogen sources like manure until the plants start to set fruit. If your soil is heavy and the rainfall is too, you probably should mulch heavily between rows and lightly (about 4 inches) between plants, putting compost or manure on your mulch as a top-dressing. If you don't mulch, loosen the soil with an iron rake when it dries, cultivate between plants to keep a crust from forming, and add some manure as a side-dressing. During dry spells, be sure to keep your okra sprinkled. But beware of overwatering.

Some okra growers say that pruning is the secret of good pod production. They claim that if you snip off one of each three leaves when the blossoms appear, pruning mostly toward the inside of branches and more heavily during prolonged wet spells, your plants will bear early and continuously up to frost and fruit conveniently, mostly on five main branches. Cutting out suckers will also give more uniform growth.

Like other members of the hibiscus family, okra can suffer from bud drop, which is aggravated by hot, dry air, sudden changes in temperature, or poor drainage. Insect pests such as the corn earworm, green stinkbug, and cabbageworm can be hand-picked in the morning, then dropped into kerosene. (To help guard your okra from stinkbugs, don't plant it near legumes, eggplants, potatoes, or sunflowers, and keep your garden well weeded. You can ward off cabbageworms by interplanting with garlic and onions.) In the South, gumbo is vulnerable to the pinkish fungus called southern blight and to wilt, which creates yellowed, droopy leaves and stunted plants. Crop rotation and deep plowing can help if your area is troubled by these fungi.

When your okra pods are from 1½ to 4 inches long, they're ready for harvesting. In fact, it's known that you can get a larger, longer yield if you gather your fast-growing pods every day or so during hot spells and/or when they're a mere three to four days old. Wear gloves—for okra can irritate the skin—and snip off the pods with about an inch of stem. Be careful, by the way, not to injure them lest they ooze and lose the mucilaginous juice that gives okra its prized ability to thicken soups and sauces.

If well treated, okra plants are prodigious yielders. Nancy Pierson Farris, a South Carolina gardener, reports that in a typical harvest season, she has "picked a quart of okra weekly from one 15-foot row."

Now, if you have a good long growing season, you can cut back your plants after a first harvest, give them a dose of manure tea, and get another crop of tasty young pods before the first frost. But you might let the pods stay on some of your plants until they mature, since you'll be compensated for any drop in yield by an extraordinary increase in the protein content of the ripening seeds. If your summer has been wet, cut off the browned pods and let them dry in a warm place before removing the matured seeds. If the weather has been dry, however, you can shell the seeds right after cutting the pods.

Young okra will keep for a day or two in a cool room if you sprinkle it with water to maintain a high humidity and provide lots of circulating air. To freeze the tender pods, cut off their stems carefully, then blanch the okra for 2 to 3 minutes in boiling water, or 5 minutes in steam. Cool and freeze whole or sliced crosswise. If you prefer, you also can freeze okra unblanched. Just slice it and coat with cornmeal. To can washed and trimmed gumbo, cover whole or sliced pods with boiling water and boil for 1 minute. Drain, reserving liquid, then pack into jars and cover with hot liquid, leaving 1 inch at the top. Then put your pints or quarts into the pressure cooker at 10 pounds pressure for 20 or 40 minutes, respectively. Okra also dries well. Just string pods that are whole, halved, or cut in rings and hang them in the open shade to dry. Or leave whole okra in the oven at 130° to 150°F. for 3 to 6 hours after blanching it for 3 minutes.

Nutritionally, immature okra pods offer good amounts of calcium—more than green zucchini, for example. Okra also contains a modest portion of vitamin A—468 international units per half a cup. The really substantial food value for this vegetable is to be found in the ripened seeds, which contain almost as much protein as soybeans, vegetable oil of good quality, an assortment of vitamins and minerals, and some fiber.

Young okra is delicious when sliced, dipped in cornmeal, and fried. It also tastes superb stir-fried with a little garlic and soy sauce, sliced and fried along with Polish sausage and tomatoes, or mixed with onions, summer squash, peppers, and tomatoes.

Young pods can be breaded or batter-dipped and fried whole for a crispy treat. Another classic dish is okra creole, in which the sliced pods are sauteed with onion, green pepper, and chopped tomatoes. As the commercial soup makers know, okra is also a natural for stews and soups since its special gumminess makes

A NEW PROTEIN POWERHOUSE

Two researchers at the University of Rhode Island, Pavlos Karakoltsidis and Spiros Constantinides, have found that although okra seeds have a lower protein content than soy, they have a more balanced amino acid pattern that makes more of the essential nutrient available to our bodies. This work becomes quite significant when you realize that an okra pod only 9 inches long can have up to 100 seeds and that field trials have shown yields of 2,000 pounds an acre are possible. Since okra can be cropped constantly until it's killed by frost, the seed makes an attractive home-grown protein.

Because the limiting factor in okra's protein profile is the sulfur-containing amino acids, okra in combination with other seed proteins like sesame or wheat flour can provide highly usable protein. Tests on baking properties showed that 5 percent okra added to whole wheat flour yielded a loaf researchers found "highly acceptable—having a distinct flavor." So if you're growing okra next year, why not dry some pods and live a little?

for a thicker, richer broth. Some folks combine okra with tomatoes to make zesty sauces. Others enjoy sliced okra boiled with black-eyed peas and drenched with lemon butter or in casseroles featuring veal or other meats. The pods can also be pickled like cucumbers with dill or hot peppers.

Use your ripe okra seeds as you would dried beans. Or you can grind them into high-protein meal or roast and grind them for use as a pleasant-tasting coffee substitute. Or use them in your favorite recipe for pickled capers. Okra leaves can be

cooked as greens, and the dried pods look attractive in autumn bouquets.

For okra seed to plant, write to Burgess Seed and Plant Co.; W. Atlee Burpee Co.; D.V. Burrell Seed Growers Co.; William Dam Seeds; DeGiorgi Co.; Farmer Seed and Nursery Co.; Glecklers Seedmen; Gurney Seed and Nursery Co.; Joseph Harris Co., Inc.; Charles C. Hart Seed Co.; H.G. Hastings Seed Co.; J.L. Hudson, Seedsman; Jackson and Perkins Co.; Le Jardin du Gourmet; J.W. Jung Seed Co.; Earl May Seed and Nursery Co.; Mellinger's, Inc.; L.L. Olds Seed Co.; Geo. W. Park Seed Co.; Redwood City Seed Co.; Seedway; R.H. Shumway Seedsman; Stokes Seeds, Inc.; and Otis S. Twilley Seed Co.

• Quite against all recommendations, okra can be planted much closer together than it normally is. We plant ours using the intensive method in beds 5 feet wide (6 feet from walkway centers) at 1-foot intervals each way, and our plants grow 7 feet tall and bear loads of fruit for freezing, pickling, and cooking. The okra stalks are woody and well rooted, so we leave them over the winter and plant climbing beans in the old okra bed. Those old stalks are well able to serve as bean poles without tipping over.

—*Frank A. Schierenberg*
Dresden, Ontario
Canada

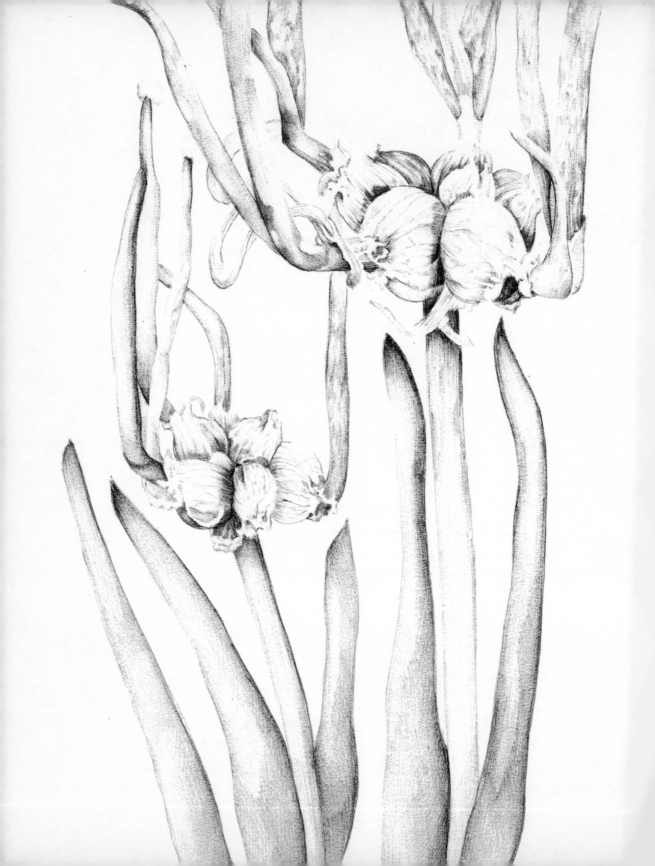

Egyptian Onion

Allium cepa, var. *viviparum*

THE honorable onion has proved itself to be an extremely useful plant ever since the origin of human history. The laborers who constructed the pyramids in Egypt were fed onions to increase their stamina and strength and before ancient Greeks and Romans clashed in battle they would dine on onions to recuperate from their usually strenuous marches. Besides being a remarkably versatile vegetable the onion is easy to grow, hardy, virtually immune to most diseases and insects, takes very little space to grow, and its flavor never fails to perk up an otherwise bland sauce or salad.

The many members of the Allium family demonstrate their strong individuality. While not often found in American gardens, the Egyptian onion may be one of the most individual of them all. This oddity of the onion world bears flowers which are followed not by seeds, but by small bulbs. The Egyptian onion, also called the Garden Rocambole or Walking Onion, sports clusters of very small, purple-skinned, pungent-flavored bulbs on top of 3-foot stalks. For this reason Egyptian onions are also called tree or top onions.

At maturity, which occurs in midsummer for onions planted early in spring, small bulbils form on top of the stalks, with a second and even a third cluster sometimes forming on top of the first. In the ground, the rounded base, which is not a true bulb, *271*

• Even here in Massachusetts, Egyptian onions can be harvested right from the garden most of the year. In winter they're dug up in thaws, or on cold days cut off at the surface with pruning shears. Frozen stalks thaw out perfectly. In zero weather with deep snow, I use the little top bulbs, storing them indoors. The stalks lengthen in late winter when they are used as scallions, or even cooked as young leeks.

—*Ruth Tirrell*
So. Weymouth, Massachusetts

Some gardeners plant their top onions in a trench 5 to 6 inches deep, then cover the bulblets with a little earth and gradually pull in more soil as the bulb grows. This method is more practical farther north to protect the bulbs from hard freezes. In any event the bulbs should be well mulched in cold northern winters.

multiplies, splitting into several sections, from each of which grows a slender scallion. The plant never makes a big onion, just long green onions about an inch in diameter. After the onion blooms, it begins to do its tricks, and becomes one of the more entertaining inhabitants of the garden. The little bulbs at the top of the stem begin to develop and then, when the big hollow stems get weak and overburdened with the weight of the bunch of little bulbs, the stems crack and buckle and the bulbils bend down and are deposited on the ground where they immediately take root and grow. A prime example of nature taking care of its own reproduction.

The self-propagating Egyption onion requires little care and will grow almost anywhere in any condition. One place it won't appreciate, however, is a swamp, since it requires good drainage. The best time to plant is from the middle of August (in the North) to November (in the mid-South) so the plants can get a good start before winter.

Although it is the least demanding of all onions, the Egyptian variety does enjoy a rich, light soil in a sunny part of the garden. The plants are heavy feeders, so for best results the ground should be carefully prepared by mixing in generous quantities of decomposed manure or compost. The hardy onions will prosper without such fuss but the stalks will be juicier and more tender if some compost has been mixed in.

Cultivation is relatively simple. The plants may be propagated from the bulbs formed in the soil or from those on the stem placed in shallow drills or beds 5 to 6 inches apart in rows separated by 10 inches. The bulbils should be at least as large as the end of your little finger to plant alone. If they are smaller, plant the whole cluster instead of breaking them apart. Plant the little bulbs about an inch deep, making sure they are firm in the soil—birds have been known to pull up loose bulbs.

After planting, you can all but ignore your top onions, except for an occasional watering and weeding. Once the little sprouts appear, begin mulching them to conserve moisture and control weeds. From this point on, watering probably won't be necessary unless the plants look too dry and begin to shrivel. A hay mulch in winter, though not a necessity for the plant's survival, will make it easier to dig up the clumps. Once the rows start to get crowded, thin them so there is a plant every foot, and use the thinnings in a salad as you would scallions.

The onions left in the row should then start multiplying by sending up new shoots and also developing bulblets on top. These new bulbs can be used to start another row or, if you're feeling generous, you can give them to friends so that they too can enjoy onions all year round. Because of their hardiness the new bottom shoots will grow well into cold weather and come up again quickly in spring. The second year you can harvest these shoots for spring, fall, and winter eating. Once the onions have established themselves, they will last indefinitely. The harvesting keeps the multiplying clumps of new shoots under control. One Massachusetts gardener has a bed of Egyptian onions which has stayed exactly the same size for over 10 years, and a gardener in Vermont has kept his bed going for more than 20 years.

The Egyptians have much better keeping qualities than other types of onion bulbs. Hung upside down in a dark place the bulbils should keep better than even garlic.

All parts of this diversified onion are edible at some time or another. The top bulbs can be used cooked or raw for flavoring. They are quite strong in flavor, so a little goes a long way. The new green tips can be used as chives, the white lower stalks can be chopped raw for a salad like regular onions, and the young stalks which were thinned can be served raw like scallions. Even the coarser stalks can be useful in adding flavor to a chicken stew. For a lively salad, the fat, hollow leaves can be gathered in spring when they are tender, slit and stuffed with cottage cheese. Egyptian bulbils are frequently pickled because they are stronger and much richer in flavor than other varieties of onions. The larger bulbils are especially good for adding to cucumber pickles.

Sources for Egyptian onions include: Casa Yerba; DeGiorgi Co.; Hemlock Hill Herb Farm (plants); Le Jardin du Gourmet (seed bulbs); Meadowbrook Herb Garden; Nichols Garden Nursery (seed bulbs); and Well-Sweep Herb Farm (plants).

• When I first saw the stems of my Egyptian onions buckle and fall over, I was distressed. Then when I discovered that this was the method they used to get their bulbils down to the ground, I marveled at the ingenious ways nature finds to accomplish its prime end of reproduction.

I use them in cooking as I choose. No one seems to know that they are any different from any other onions—but I think of them as quite strong in flavor. They are too small to bother with for boiled onions, and their outer skins are tough and somewhat bothersome to peel off. They do make a good source for small bits of onion to fry for flavoring.

—*Catharine O. Foster*
Bennington, Vermont

Welsh Onion

Allium fistulosum

THE name given to the Welsh onion joins Jerusalem artichoke and asparagus pea as another case of misleading nomenclature. Welsh onions have nothing to do with Wales, but seem instead to have everything to do with a German word, *welshe,* meaning "foreign." This mild-flavored, bulbless onion is believed to have originated in Siberia or eastern Asia—quite a distance from Wales. History tells us that the German word was first applied to the onion when it was introduced into Europe towards the close of the Middle Ages. It is also called the Japanese Bunching Onion, the Evergreen Bunching Onion, and the ciboule.

Today, Welsh onions are grown around the world, particularly in the East. In China and Japan, this hardy vegetable has been cultivated since prehistoric times, and it is today considered a principal garden crop. In Britain, this onion has been part of home gardens since it made its debut there in 1629. But despite the Welsh onion's popularity in some parts of the world, it never really caught on in the West, and many gardeners here in the United States have never heard of it.

Still, it is well suited for American gardens, and has a great deal to recommend it. It is extremely hardy, and can be left outdoors and pulled as needed all winter long where winters aren't too severe. In colder climates, Welsh onions can be dug up and transplanted to the cold frame or greenhouse for the winter, without suffering any ill effects. The plant is a perennial, and once established from seed, these onions are easy to grow. Finally, their flavor is quite mild and sweet—an asset to many dishes.

Like the leek, the Welsh onion never forms a true bulb. Its roots are elongated, tapering, and slightly swollen with strong fibers, and its stems and leaves are hollow. During the second year of growth, the plant sends up a flowering stem which can grow as tall as 20 inches, and which bears a globular head of yellowish white flowers. Unlike leeks, however, the Welsh onion multiplies at its base, much as shallots do.

Cultivation of this mild-mannered onion is somewhat similar to that of the leek (in fact, in Japan it is wrongly referred to as the "Japanese leek"). Special attention must be given to ensure tender, long, white stems. The trench method recommended for leeks (page 229) is also a good way to grow Welsh onions. The base of the bulb, if planted 6 or 8 inches down in loose soil, will provide the long, delectable stems you are looking for. The bulbs are planted in the trench during their second year of growth. To start the onions from seed, they are planted in conventional rows or beds. The Welsh onion will usually thrive in an ordinary soil. Simply plant the seed about ¼ inch deep in good, friable soil. Although you can plant them almost any time in milder climates, it is best to plant in spring where winters are harsh. The plants should grow to a fairly mature size the first year, with some multiplying taking place at the base. By the second year you will have an abundance of mature Welsh onions. When the second year of growth begins, dig them up and transplant them into a 6-inch trench which has been manured beforehand. Then, like you would for leeks, fill in the trench so that the bottom part of the onion is covered.

Like all onions, the Welsh variety suffers if the soil dries out. To avoid this, heap up a thick mulch of compost around them. This cover will not smother the onions as a heap of soil might. Also, mulching makes it easier to dig up the clumps with a trowel. Remember that mulches need continual renewal to make sure that they do not break down.

• Here in California, we leave Welsh onions in the ground through the winter. We pull them for use in cooking all winter long, so you can see that our mild cold does not cause them to deteriorate in quality. In climates where the ground freezes they should be moved into a greenhouse or cold frame. One of the virtues of this onion is that it can be transplanted and moved around from year to year without a serious setback. When pressed for garden space, I've put them in a tub or an odd corner of the garden to maintain life until I could find a better spot for them. They are, once established from seed, very easy to grow. Welsh onions will stay alive in even poor soil, but they become sweeter and more tender if the soil is richer.

—John Meeker
Gilroy, California

By the end of summer, the onions will be ready to harvest. Dig the clumps as you need them, and there won't be any need to worry about storage.

For a family of four who uses green onions principally for seasoning, a small clump 1 foot by 2 feet will be sufficient. On the other hand, if your family fancies Chinese or Japanese dishes you should probably double the size of that clump.

Welsh onions are most often used like green onions, to furnish young growths for salads, stews and soups. Their mild flavor is a welcome addition to all your favorites. Try Welsh onions in homemade potato soup for a real treat. These mild onions combined with fried, crumbled bacon and grated sharp cheese make a wonderfully zesty flavoring for quiches and omelets. During the second year of growth Welsh onions produce seed stalks which, if cut soon enough, are excellent to add flavor to your cooking. They are used much the same way as chives. In Japanese and Chinese cuisine both the "bulbs" and stalks are a standard ingredient in stir-frys and soups, in combination with meats and/or other vegetables.

There are several varieties of Welsh onions, all very similar in appearance and taste. The NEBUKA (HE-SHI-KO or JAPANESE BUNCHING) is frequently grown commercially as a green onion. It is hardy and resistant to pink root and onion smut. The EVERGREEN onion is a hardy multiplier whose long white stems are most desirable for cooking. Other similar varieties include the EVERGREEN LONG WHITE BUNCH-ING SCALLION, SAKATA'S EVERGREEN HARDY WHITE BUNCHING ONION and the PROLIFIC WHITE BUNCH-ING ONION.

Seeds are obtainable from many companies, including W. Atlee Burpee Co.; Comstock, Ferre and Co.; William Dam Seeds; DeGiorgi Co.; Joseph Harris Co., Inc.; J.L. Hudson, Seedsman; Jackson and Perkins Co.; Johnny's Selected Seeds; Meadow-brook Herb Garden; Nichols Garden Nursery; Geo. W. Park Seed Co.; Redwood City Seed Co.; R.H. Shumway Seedsman; Sunrise Enterprises; and Otis S. Twilley Seed Co.

Orach

Atriplex hortensis

SEVERAL members of the goosefoot family go by the name orach, the most common edible species being garden orach (*Atriplex hortensis*). Known also as salt bush, musk weed, mountain spinach, French spinach, and butter leaves, it is a tall, hardy annual with coarse leaves that can be cooked and eaten like spinach. Along with the other, closely related species of orach, garden orach shares a long history as one of the oldest cultivated plants. A native of Europe and Siberia, it was widely recognized among the ancient Greeks and Romans who used it to soothe sore throats, ease indigestion, and cure jaundice. During medieval times, orach became very popular in European countries where it was used as a culinary as well as a medicinal herb. Italians added it to their pasta and the French and English used it in various soups and stews or served it steamed as a side dish. Early settlers brought orach to the New World where it became a fairly standard vegetable in seventeenth- and eighteenth-century gardens. For some reason, however, it fell out of favor during the nineteenth century and, although it continues to be grown throughout Europe, has been replaced by spinach in this country.

As a hardy, drought-resistant green which thrives in alkaline soils, orach deserves reconsideration on the part of American gardeners. In the Great Plains and Mountain states, where spinach shoots to seed before developing good-sized leaves, orach flourishes. A much hardier crop than New Zealand spinach, it is an excellent choice for such areas. With several successive plantings made early in the season, a supply of tender young orach leaves can be harvested continuously throughout the growing season.

There are three main types available: white orach, which has pale green leaves; red orach, which has dark, reddish stems and leaves that turn green when cooked; and green orach. The white type is generally considered the sweetest and most tender and, as such, is the most often cultivated on a large scale. The other types are grown primarily as edible ornamentals in home gardens. Assorted varieties can be purchased which range in size from 4 to 10 feet in height, the medium-sized ones being easier to harvest and generally the most desirable. All forms have long-stalked, triangular or shield-shaped leaves which attain lengths of 3 to 7 inches. The yellowish flowers appear in early or midsummer. Orach is monoecious, meaning that both male and female flowers are borne on the same plant.

Seeds can be sown as soon as the ground is workable in the spring. Like spinach, this vegetable prefers a shady, well-drained location which is fairly fertile. Work some manure or compost into the soil before planting, then side-dress as needed during the season. Orach is extremely tolerant of alkaline and saline soils.

Plant seeds thinly in rows 15 to 24 inches apart. Water gently and frequently until the seeds germinate. When the young plants have become established, thin them to stand 8 to 12 inches apart within the row. Don't count on replanting the thinned seedlings, as orach has very fragile roots and does not take transplanting very well.

Although this is a very drought-resistant crop, the highest quality greens are obtained when the plants have been watered freely so as to maintain rapid growth. A thick mulch of straw or compost will help control moisture as well as improve sandy, saline soils. It's a good idea to control weeds during the early stages of plant growth, but weeding quickly becomes unnecessary as the plants reach heights of several feet and develop the large leaves that shade the bed.

The tender leaves which arise from the top of the plant are considered the best to eat. Older leaves are very tough and make for fairly unpleasant eating. The leaves of flowering plants may also be less desirable than those of immature ones. Most gardeners pinch back the flowers as they appear, to encourage leaf growth and branching. In this way, edible leaves can be picked from a single planting of orach all summer long. Successive sowings every two weeks from spring to midsummer will guarantee a continuous supply of even higher quality orach.

To harvest, simply break off the young leaves and stalks. Several side branches will arise in just a few days, bearing more tender new leaves. Keep the plants well-trimmed, eliminating any diseased or insect-infested leaves as they appear.

Young stems and stalks can be used in all of the ways spinach or sorrel is used. Orach has a mild flavor and contains much less acid than most other types of spinach. Because they grow well up off the ground, the leaves are quite clean and free of insect damage. Boiled and buttered, creamed, added to quiches, rolled up in crepes, tossed in a cold salad, or simply added to the soup pot, orach is not only a fine spinach substitute, but a delicious vegetable in its own right.

Orach seeds are available from J.L. Hudson, Seedsman; Meadowbrook Herb Garden; and Redwood City Seed Co. Well-Sweep Herb Farm supplies plants.

Hamburg Parsley

Petroselinum crispum, var. *tuberosum; P. hortense,*
var. *radicatum;* or, *P. sativum,* var. *tuberosum.*

AMERICAN seed companies have only recently begun
to offer this versatile Mediterranean vegetable. Known also as
Dutch or turnip-rooted parsley, it is widely cultivated throughout Europe where it is used in soups and stews, and roasted with
meats. In the kitchen, the most important part of the plant is the
smooth-skinned, fleshy white root, which resembles a short
slender parsnip, but tastes more like celeriac. Top growth
consists of a mass of flat, parsley-like leaves. These, too, are
edible, and can be used like curly or Italian leaf parsley, as a
flavoring and garnish.

As with most root crops, seeds for Hamburg parsley should
not be planted in a recently manured bed. A soil too rich in
nitrogen causes forked roots and excessive top growth. For
straight, fat roots, the soil should be deeply cultivated, manured,
and limed if very acid, the autumn before planting.

If you forget to prepare in advance, there's no cause for
alarm—this vegetable will tolerate very poor land and partly
shady locations as long as sufficient moisture is provided. Soils
too weak to support other root vegetables will often produce
perfectly adequate crops of this very hardy parsley.

Like its leafy parsley cousins, this type is a cool-season plant
which grows best when temperatures are between 50 and 70°F.
The highest percentage of seed germination is obtained when

temperatures are between 55 and 65°F. When the mercury dips to below 45°F., plant growth slows considerably. Although Hamburg parsley is very resistant to frost, freezing temperatures slow its growth to a near-halt. Similarly, in temperatures above 75°F., growth is also slowed, and the plant will surely bolt the second year if the roots haven't been dug.

For even, continuous development of Hamburg parsley, sow seeds as soon as the ground can be worked in spring, or in late autumn if you live in a mild-winter region. Where summers are not terribly hot, seeds may be planted in July to winter over and produce an early crop the following spring. Hamburg parsley requires a long growing season; those seeds sown in March will probably not be ready to harvest until October or November. If your growing season is short, starting the plants indoors early will allow the roots to fully develop before the onset of cold weather. Plant seeds thinly, ¼ inch deep in rows 12 to 18 inches apart. After several weeks, thin the seedlings to stand 6 to 9 inches apart. Since germination is very slow, some gardeners recommend soaking seeds overnight and irrigating the soil before planting. Mixing seeds with radish seed helps to mark the rows. For efficient use of space, interplant your parsley root with a small lettuce variety, radishes or some other short-season crop. Hamburg parsley may also be planted as a border if it is in full sun.

Water the garden bed frequently and gently until the seedlings are well established. Later, when the roots are deeper and stronger, less frequent watering will do the trick. A mulch of compost or straw will eliminate the need for a lot of irrigation and, of course, helps to control the weeds. Occasional light applications of wood ashes or granite dust, while not absolutely necessary, will further encourage root development and will also help to repel snails and slugs.

Like carrots, Hamburg parsley can be harvested at any time after the roots reach a reasonable size. Medium-sized roots of 5 to 7 inches in length are much more tender and sweet than the larger, thicker ones. As with parsnips and salsify, roots which have undergone at least one frost usually have a higher sugar content than those harvested during summer months. By all means delay at least part of the harvest until after frost, for best flavor. Where winters are especially severe, it may be necessary to lift most of the roots in November for winter cold-storage in damp sand in a root cellar or cool basement. Elsewhere, just cover the plants with a protective straw or leaf mulch and dig roots as needed throughout the winter. The flavor of these freshly dug parsley roots is much superior to that of the stored ones.

Harvested roots should be scrubbed before cooking, or parboiled and peeled to remove the slightly hairy skin. Avoid peeling the roots before cooking, as they are likely to discolor. The flavor of these small but succulent roots has been likened to a mild parsnip, or more accurately perhaps, parsley-flavored celeriac. Hamburg parsley can be mashed like parsnips, sliced and fried, grated into salads, or roasted with a joint of beef. They are very good French-fried; or mashed, mixed with salt, pepper, nutmeg, beaten egg, and bread crumbs, and made into patties that are fried in oil. They can be batter-fried, too, or even glazed. Served alone, four to six roots usually make an average-sized serving. Parsley root can be prepared in all of the ways you fix celeriac, and when sliced or grated, makes a savory addition to the soup or stew pot. In fact, the plant's foliage is such a popular soup and stew ingredient that it has earned the name of "soup greens."

Seeds of Hamburg parsley are available from many suppliers, including W. Atlee Burpee Co.; Casa Yerba; Comstock, Ferre and Co.; William Dam Seeds; DeGiorgi Co.; Farmer Seed and Nursery Co.; Gurney Seed and Nursery Co.; Joseph Harris Co., Inc.; Charles C. Hart Seed Co.; J.L. Hudson, Seedsman; Johnny's Selected Seeds; J.W. Jung Seed Co.; Mellinger's, Inc.; Meadowbrook Herb Garden; Nichols Garden Nursery; L.L. Olds Seed Co.; Geo. W. Park Seed Co.; R.H. Shumway Seedsman; and Stokes Seeds, Inc.

Asparagus Pea

Lotus tetragonolobus or *Psophocarpus tetragonolobus purpurea*

NEITHER a genuine pea nor a substitute for asparagus, this unusual vegetable is said by gourmets to combine the flavors of both, and its unique taste has inspired both raves and rejections. To the sympathetic palate, asparagus peas are an incomparable delight, and are served simply, with melted butter, to be savored fully. They are practically unknown in this country, and for gardeners who discover a taste for them either by visiting Europe or by experimenting with new crops, these "peas" become a yearly fixture in the garden.

There are two different botanical classifications for the asparagus pea, both of which seem to appear with about equal frequency. Whether they represent two distinct plants or just two names for the same one is difficult indeed to discern. In any event, both are grown and eaten the same way, so the difference in nomenclature may be of little consequence for our purposes.

Most ethnobotanists place the origin of the asparagus pea in southern Europe, although others say it is native to India. Today it is grown primarily in Europe and Asia. The plants are low growers, with trailing stems reaching about a foot and a half in length. With its delicate light green foliage and its unusual reddish brown or purplish flowers, the asparagus pea dresses up any garden; in fact, it's pretty enough to grow in the flower bed.

Seed is sown in spring, after the soil has warmed and all

287

danger of frost is past. Pick a spot that gets full sun. Plant the seeds about 3 inches apart, in rows 12 to 18 inches apart. The plants are quite sensitive to cold, so if you live in a cool zone, you may want to start them indoors. The seedlings should be set out as soon as possible after the ground is warm, but be careful not to damage their delicate roots. Rich, fertile soil is most often recommended for asparagus peas, although Linda Bayliss of Michigan reports that hers "flourish and even self-seed in soil that is little better than hard clay." One type of soil the plants don't do well in is sand. If you want to take out insurance on your crop, prepare the soil with wood ashes or other potash-rich material, and side-dress four weeks after planting. If you make a practice of inoculating all your legumes, save it for your other peas and beans. Since the asparagus pea is not a true legume, commercial inoculants do not affect its seed.

The plant's lovely flowers are followed by curious-looking pods sporting four longitudinal ribs or "wings" which have given rise to two of the vegetable's nicknames—winged pea and winged bean. In some quarters it is also known as the goa bean or Manila bean. Asparagus peas mature in 90 days, but the pods are best when eaten young. To catch your asparagus peas in their prime, pick them when they're only an inch long; older pods become tough. If you keep up with the harvest, the plants will continue to produce for a long time, and stand up very well in the heat when other peas have wilted. A mulch of hay or other material will help the plants to hold up better.

Asparagus peas are easy to prepare and serve. To enjoy this gourmet treat, simply steam the whole pods until tender, about 10 minutes, and serve with melted butter poured over them. If the crop has gotten away from you and the pods have reached their mature length of 2 to 3 inches, don't despair. The older pods develop a tough inner core, but the outer pod is just as tender and delicious as the young peas, although not quite as daintily eaten. To prepare these older pods, steam them for 15 to 20 minutes, drain, and serve with melted butter alongside. They are eaten by dipping, one at a time, into the butter, sucking off the outer pod, and discarding the tough center. It's rather like eating artichokes.

Though the asparagus pea is valued primarily for its pods, other parts of the plant can be eaten as well. Young leaves and shoots lend a distinctive flavor to soups, stews, and vegetable curries, and the ripe seeds may be roasted and eaten with rice. The small tubers formed underground are said to adapt themselves to many standard potato recipes, and can be eaten raw as well.

Seed for asparagus peas is available from three companies: J.L. Hudson, Seedsman handles *Lotus tetragonolobus;* Geo. W. Park Seed Co. and Thompson and Morgan, Inc. carry *Psophocarpus tetragonolobus.*

Sugar Pea

Pisum sativum, var. *macrocarpum*

THIS edible-podded legume so often featured in fine Chinese dishes was not developed until the nineteenth century and then on English, not oriental, soil. But the Chinese recognized a good thing when they saw it and so extended its culinary uses that the English sugar pea soon became known as the Chinese snow pea.

Snow peas are very easy to grow and especially sweet-tasting when freshly picked and eaten. With fresh-market prices up as high as four or even five dollars a pound, they should be a first choice for the home garden. Fast-growing, cool weather vegetables, they are popular for their delicate and sweetly scented flowers as well as their light green pods. Since the pods lack the parchment-like inner lining of garden pea pods, they can be eaten whole and remain sweet and tender until quite large.

Several varieties are available in the United States. The white-flowered MAMMOTH MELTING SUGAR pea is a tall, climbing strain that produces many high-quality pods. Pods of the DWARF GREY SUGAR pea are smaller and preferred by some gardeners. Plants of this variety mature some 15 days before climbers and grow to only 3 feet in height.

In some northern coastal or mountain regions, snow peas can be grown continuously from early spring through fall. In very mild regions, they will thrive as a winter and early spring vegetable. Elsewhere, where summers are fairly hot and winters cold, snow peas can only be grown as an early spring and fall crop.

Before planting, prepare the bed with compost or well-rotted manure. This will not only improve the moisture-holding capacity of the soil, but also provide plants with enough nitrogen to establish themselves. No other nitrogen fertilizer is usually necessary since, like other legumes, snow peas benefit from atmospheric nitrogen which bacteria fix in the plant roots. Add ground limestone, wood ashes, and ground phosphate as necessary to establish a slightly acid soil, with a pH of 6 to 6.5, that is rich in phosphorus and potassium. These elements encourage root development and setting of flowers.

Like all peas, these varieties should be planted very early so that the crop will mature before the summer heat arrives. For fast germination, soak seeds in warm water overnight. They benefit greatly from inoculation with a nitrogen-fixing bacterium. One California gardener reports that he soaks his seeds overnight, spreads them on a damp towel, rolls up the towel and stores it in a warm, dark place until the seeds show signs of sprouting. Whether you sprout your seeds first or not, sow the presoaked, inoculated seeds 1 to 2 inches deep. Dwarf varieties should be spaced at 2- to 3-inch intervals in rows 18 to 24 inches apart. Climbers are traditionally planted 2 inches apart in double rows 3 inches apart, with a trellis or wire support between the rows. Some gardeners prefer to sow the seed very thickly, and thin plants when they are about 1 inch high. One package of seed will do for a 15-foot row, one pound for 75 to 100 feet. The object is to make the stand of sugar peas thick, but not overly crowded. Planted closely together, the tall varieties climb over and support one another.

Culture of snow peas differs little from that of common garden peas. Water the plants thoroughly and deeply during dry spells, especially during the early stages of pod formation. As the season progresses, cultivate deeply between rows and shallowly near the developing root systems of plants. Frequent cultivation encourages rapid growth and helps control several insects and diseases which may threaten the stand. Early plantings made while the ground is still cold should not be mulched, but later in the season, plants will benefit from a layer of sawdust or straw. Tall varieties should be provided with a sturdy support of tree brush, chicken wire, or strings before they are 3 inches high. Unstaked plants are not only hard to pick, but also seem to produce smaller yields.

Check the plants regularly for insect and disease damage. Aphids find snow peas attractive, but natural predators, aromatic interplantings, and an occasional onion or garlic spray should control them. Root rot is not usually a problem where good drainage is provided, but if it does appear, yearly rotation eliminates it. Oscochyta blight is another disease that infects sugar peas. Grey blotches covered with small brown spots appear on the pods, and the plant stems develop sunken, rotting regions. Leaves may also turn grey as the plant weakens and eventually dies. If blight strikes your pea patch, eliminate any diseased leaves or plants and burn them. To avoid the problem the following year, plant the peas in a new location.

When grown in a well-drained soil, given plenty of moisture, and properly fertilized, Chinese peas are generally very resistant to such diseases and are prolific, rapid growers. They can produce almost double the harvest of garden peas if treated well. If successive sowings are made every 10 days from early March through early May, fresh peas can be picked from early May right through the first few weeks in July. Plantings made in mid or late August supply peas throughout the early fall months. Unlike garden peas, sugar types can withstand some frost once the flowers have set. Indeed, it is possible to pick these "snow peas" while the season's first, or perhaps last, flurries are falling.

Most varieties require 60 to 70 days from seeding to maturity, at which time the green pods are fully developed but the peas within are not yet formed. Most snow peas supply two harvests, the main crop picked every other day for one or two weeks, and the second, smaller crop harvested soon after. Immediately after this second harvest, the vines should be turned under or fed to livestock. They are, of course, very rich in nitrogen and are also excellent enrichers of the compost heap.

To harvest edible-pod peas, pick them during cool evening

• . . . I want to pass along the advice of a Chinese chef for whom I once worked. He insisted the absolutely best snow pea was one grown in cool weather in the home garden, and picked at a point just after the peas begin to swell in the pod. The only way to determine this point is to taste the peas daily. Sooner or later, you'll bite into one and it'll have an exquisite sweet crisp taste. Notice its general appearance, and pick all your peas as they reach that stage. This is the point where the pea has accumulated the peak amount of sugar. The pod is still tender, and the gently swelling peas add a special touch. Immediately afterward the sugars will begin to turn to starch and the pods will get tough.

—*James Jankowiak*
Eureka, California

or early morning hours and serve or freeze them immediately. Dwarf pods should be 1 to 3 inches long, and climbers 3 or 4 inches. These larger ones are usually cut into bite-sized pieces before cooking. They need no other preparation beyond pinching off the far end with your fingernails. If you've let the crop go a little too long before picking, and the peas are beginning to swell noticeably in the pods, it's a good idea to break the stem end from bottom to top and peel away the strand of tough fiber.

Cook fresh snow peas within an hour of picking to preserve all of their natural sugars. They are delicious served raw in salads, added to innumerable Chinese, Japanese and Polynesian dishes, stir-fried alone or with other vegetables (they combine especially well with sliced water chestnuts or jicama) and seasoned with soy sauce, or simply steamed and served with butter and a touch of salt. Snow peas have much to recommend them: there is no waste in preparing these peas, they are one of the first and most delicious spring vegetables, and they have a high vitamin A and C content to boot.

Seeds for sugar peas can be purchased from just about any seed house.

• With respect to freezing snow peas, I would like to suggest the following method. I fry enough for a meal in butter for a few minutes until they change color from light green to darker and shiny, package and freeze. To serve, simply thaw and fry until heated through.

—*Valerie J. Pavlovski*
Palgrave, Ontario
Canada

Peanut

Arachis hypogaea

ACTUALLY a legume, the peanut is widely known and wildly loved as a nut—the most popular one in the world! This irresistible snack has been found in Peruvian ruins from 800 B.C., and has even been identified as a motif in ancient South American pottery. In the sixteenth century, the Portuguese carried the peanut from Brazil to east Africa, and the Spanish introduced the tropical crop to the Philippines. A century later, slaves took it from Africa to Georgia and the Carolinas.

Today the peanut is a most important crop all over the tropics and subtropics, and vast amounts are grown in India and China. In the United States, peanuts are now our fifth largest commodity. Grown commercially as far north as Virginia, they have also prospered in gardens as far from Dixie as Illinois, New York, and Massachussetts. Not known in its wild form, the peanut is also called the groundnut or earth almond, and is known as the pinda or pinder in the West Indies and parts of the South. The slang name *goober* (an African word for groundnut) found its way into a rollicking Civil War song in which a Confederate general who thinks he hears Yankee rifles "turns around in wonder, and what d'you think he sees? The Tennessee Militia eating goober peas!"

Growing from 12 to 24 inches tall, peanut plants may be erect or creeping. The attractive vines are bright green and have showy yellow pea-like flowers, which are sterile, and inconspicuous fertile flowers that grow in the lower leaf axils. After these lower flowers are fertilized, they send out long peduncles. Usually called pegs or runners, these stalks bend over to bury their tips in the soil. The peanuts—their seeds—form underground on the tips of the runners.

The many peanut cultivars may be divided into the Virginia, Spanish, and Valencia types. Virginia peanuts come in both erect and creeping varieties with one or two seeds to a pod—seeds that have a 30-day dormancy period. Also yielding two nuts per pod, Spanish plants produce smaller peanuts with no dormancy period, and their vines are always erect. If you've ever bought roasted peanuts that have three or four nuts in a shell that doesn't narrow between the peanuts, you've sampled the Valencia peanut. Without a dormancy period, Valencia types grow on upright plants.

If you're growing your peanuts in a region with a long hot season, you might want to try VIRGINIA BUNCH, a variety that will give you one or two big light brown nuts per pod. Northern gardeners claim good luck with RED TENNESSEE, a Spanish type that has two or three small red-skinned nuts per small pod. With upright central foliage and semi-reclining outer branches, this variety is about 2 feet high and can be urged more erect to catch more sunlight and ripen faster and more completely if you plant it 1 foot apart in rows 1 foot apart. Red Tennessee generates a high-yielding, easily harvested underground cluster around the center stem (as opposed to the more scattered harvest you get with creeping varieties, which spread widely and grow peanuts at the runner nodes).

If the shells are thin, you can plant peanuts in their hulls. Or the kernels can be sown shelled. In the South, plant your seed peanuts 4 inches deep; in the North, about 1½ inches deep. Since the seeds can rot if you plant too deep or if there is heavy spring rain, it is extremely important to make sure the soil is well-drained. If mounds are your style, keep them 18 inches apart and plant four nuts in each. Keep rows 2 feet apart and space the seeds as little as 3 inches apart since some of them probably won't germinate. Thin to 1 foot between plants.

Though southerners plant in spring after the soil has had a chance to warm up some, folks in the North sow at about the last frost date to stretch out the growing season, for peanuts need about four to five months of fairly hot weather and good, even rainfall to yield bumper crops.

Like other legumes, the peanut is not too fussy about soil in terms of richness, though well-worked-in organic fertilizers can help. It's essential, however, that the soil be acid (pH of 5 to 6) and very loose so the drooping pegs can penetrate it easily and start forming pods quickly. In the North, your crop will have its best chance in warm sandy soil in a sheltered place with southern exposure. For finest results you shouldn't plant peanuts in the same place more than once every three or four years. Try rotating them with root and leaf vegetables and in large plantings, with soil-building crops like clover.

Peanut vines are hardy and bothered by few insects or diseases. Your biggest problem is likely to be weeds, and you can uproot these and provide desirable aeration by cultivating occasionally from the time the plants are 6 inches high until they flower. When the vines are a foot high, hill soil up around them as you would with potatoes. This will help the fruiting pegs to bury their tips sooner and to start forming seeds more quickly. After the plants are hilled, mulch between the rows with straw.

You can also grow peanuts in the greenhouse, planting them 3 to 6 inches apart if shelled or 8 inches apart if unshelled. Cover with 1 inch of soil. Another possibility is to hunt up a large pot and display a good-looking peanut vine as a house-plant. It will provide lots of fun for children and guests. Just make sure that in return *you* provide 16 to 18 hours of bright light every day.

Your goobers will be ready to harvest when the foliage begins to yellow slightly, when sample nuts look fully grown, and when the inside of the pods begins to color and to show darkened veins. Actually, it's best to leave peanuts undug until mid or even late October, for food stored in the stems will feed the nuts even if the leaves are killed earlier by frost. This delayed harvesting is especially important in areas with shorter

Peanut Bread

One unusual and delicious way to make use of your peanut crop is to make a big loaf of peanut bread. Sprinkle two packages of yeast over 1¾ cups of warm water. When the yeast is dissolved, add 3 tablespoons honey, 1½ teaspoons salt, ¼ cup instant nonfat dry milk powder, and 1½ cups peanut flour (made by finely grinding raw peanuts). Add 4½ cups of whole wheat flour—or enough to form a moderately stiff dough. Then turn the dough out onto a wooden board and knead until it's smooth and elastic—about 7 minutes. Place the dough in a large, oiled bowl, cover, and allow to rise until double in bulk. Then punch it down, shape into one loaf, and place in a large oiled bread pan. Cover and allow to rise until double in bulk. Bake at 350°F. for about 50 minutes.

Peanuts + Petunias = A Beautiful Crop on Improved Soil

Last spring we needed to redeem a bad driveway situation created by a recently finished concrete block fence and an intervening 3 feet of bare earth—too unsightly to be left alone for the summer. We decided on peanuts because we knew their pretty foliage would make good ground cover and we needed the soil-enriching nitrogen produced by the dense root system. For a companion, we picked the rose-tinted BAL-CONY ROSE petunia because it is hardy and requires very little attention. In a few weeks we had a pretty 3-foot border bed—the kind gardeners dream about. Here's how we did it.

For a mass effect, we allowed about 2½ feet between the peanut seeds at planting time to provide room for spreading. We followed a hit-and-miss pattern because we wanted definitely to avoid

growing seasons, since the well-formed pods you eye so proudly in September will still be mostly empty. So be patient.

When you do harvest, lift each bush carefully with a garden fork and shake the dirt off, checking the hole for any pods that may have dropped off. You can dry the whole plants by hanging them in a well-ventilated attic or outbuilding in the North or in an open shed in the South. Or you can pluck the pods from the roots immediately and spread them in flats or on wire screening to dry in a warm dry place. (Don't use a basement or other cool, damp area since the moisture in the uncured pods will cause them to mildew and rot.) After two months of curing time your peanuts will be ready to be enjoyed raw or to be shelled and roasted for 15 to 20 minutes in a 300°F. oven, until lightly browned. Shelled raw or roasted, peanuts will keep well for a few months in the refrigerator if you store them in tightly closed jars or in sealed plastic bags. For longer storage, you can use airtight containers or sealed plastic freezer bags and freeze them for up to two years.

When you harvest and shell your peanuts, save the vines and shells to till back into the garden in fall along with leaves. This decaying litter will help give you well-oxygenated soil the next summer, and heavy rains will change it to compost by the next summer's end—just in time to furnish needed nutrients for the next generation of peanut plants when their nuts are forming.

Rich in vitamins B and E, peanuts contain up to 30 percent protein. They also offer a 40 to 50 percent content of healthful oil that has a better than 3 to 1 ratio of unsaturated to saturated fatty acids. One way to enjoy this high-energy food is to run equal amounts of roasted peanuts and raisins through a hand-cranked meat chopper several times. The delectable sugar-free candy that results is called gorp, and the demand for it just might use up every peanut on hand. Unlikely excesses of peanuts that don't find their way into cookies, ice cream, tossed salads, or even Southern-style peanut soup can be made into peanut butter. Just spin 2 cups of roasted peanuts in the blender, then gradually add ¼ cup of oil, 1 tablespoon at a time, blending each addition. Turn the blender off from time to time and use a knife or narrow rubber spatula to push the peanuts away from the sides and blades of the blender. Add ¼ teaspoon of salt, blend once more, and then store your 1½ cups of peanut butter in the refrigerator in a covered jar.

To get the maximum food value from your peanuts or peanut butter, you might want to put to work some of the fascinating research results given in Frances Moore Lappé's classic book, *Diet For A Small Planet.* As this extraordinary guide to economical, ecological eating explains, by eating peanuts in combination with certain other vegetable or low-cost animal proteins, you can actually increase the quality and quantity of protein in the foods used. So, if you serve ⅞ cup of peanuts (or ½ cup of peanut butter) at the same meal with ¾ cup of liquid skim milk (or 3½ tablespoons of instant nonfat milk powder), you are creating usable protein equal to that in over 5 ounces of steak! Similarly, the above amount of peanuts or peanut butter served together with 2 tablespoons of noninstant dry milk powder or soy flour and a little less than 2 cups of whole wheat flour will yield as much usable protein as over 7 ounces of steak. You can also amplify peanut protein by enjoying roasted peanuts and sunflower seeds together in grated carrot or waldorf salads—or in any other way you like. So put your peanut power to work by preparing breads, cookies, salads, and main dishes that will give you top-quality protein for pennies a serving.

If you plan to grow peanuts in the North, you might want to get some seed from a neighbor and then save some of your own crop for seed, for peanuts grown in the North will give you better results there. (Do remember, though, that roasted peanuts won't germinate!) Peanuts for planting can also be purchased from Burgess Seed and Plant Co.; W. Atlee Burpee Co.; Comstock, Ferre and Co.; William Dam Seeds; DeGiorgi Co.; Farmer Seed and Nursery Co.; H.G. Hastings Co.; J.L. Hudson, Seedsman; J.W. Jung Seed Co.; Lakeland Nurseries; Earl May Seed and Nursery Co.; Mellinger's, Inc.; Nichols Garden Nursery; L.L. Olds Seed Co.; Geo. W. Park Seed Co.; Redwood City Seed Co.; R.H. Shumway Seedsman; Spring Hill Nurseries; Stokes Seeds, Inc.; and Thompson and Morgan, Inc.

rows of plants set in a formal design.

Before planting, the peanuts were soaked for about an hour in just enough water to cover them. Then with a small hand tool the ground was worked only at the spot where the seed was planted—at a depth three times its size. Next, the entire lot was well watered, and in two or three days the peanut vines came up to show their fan-like leaves and to cover the area with a lacy green foliage.

The peanuts received no other attention except for frequent watering. When the delicate yellow blossoms appeared underneath the vines, they produced sprouts which turned down into the ground and buried themselves to grow into peanuts harvested later in the summer after the vines turned yellow. A large mass of roots formed on each plant to produce clusters of nitrogen-rich nodes which enriched the raw soil.

—*Thelma Bell*
Morehead, Kentucky

Popcorn

Zea mays var. *everta*

CORN has been grown by man since prehistoric times, and perhaps the most primitive of all varieties of this staple food is popcorn. It's been with us for so long, in fact, that archaeologists believe that popping may have been the first way ancient man made use of corn. In Peru, utensils were found that the Indians are believed to have used for popping corn thousands of years ago. In the Bat Cave in New Mexico, actual popped kernels of corn thousands of years old were discovered.

Popcorn was introduced to the white man at the first Thanksgiving dinner in 1630, when the brother of Chief Massasoit brought a deerskin bag filled with popcorn as his contribution to the dinner.

Much as the Indians and colonists enjoyed the popcorn of their time, it was far inferior to the tender, tasty varieties we enjoy today. Today's popcorn comes in several varieties, all of which are hybrids, and all of which turn white when they pop. The most unusual is strawberry popcorn, with its deep, mahogany red kernels set in small, round, 2-inch-long ears. Other colors that popcorn comes dressed in are yellow, black and white.

The cultivation of popcorn does not differ from that of ordinary corn. The same cautions apply to popcorn as to other corn varieties: do not grow it near other corn, particularly sweet corn. Popcorn is an open pollinated corn and it will dominate any sweet corn pollen it competes with. Sweet corn grown close to popcorn will fill out poorly and with a great number of deformities. Also, it's best to buy your seed corn fresh each year. Because today's popcorn is all hybrid varieties, seed saved from the harvest and planted the next year may revert to an earlier strain and produce a smaller yield.

Just like other types of corn, popcorn is a heavy feeder. For best results, the soil needs advance preparation, fertilization during growth, and yearly crop rotation. Apply a lot of manure in the fall to the area where corn is to be grown. It will break down during winter and be ready by spring. Another method is to lay down compost or other organic fertilizers such as fish and blood meal, fertilizers high in nitrogen, and dig in the material as soon as the ground can be worked.

A rule for sweet corn is to plant when the soil has warmed up to above 50°F. The same rule can be applied to popcorn, for all varieties of corn come from the warmest regions of the Americas.

Put the seeds in loose ground no less than an inch, and in tight soils no deeper than an inch. Space them about 6 inches apart in a bed and thin out the weaker plants later to about 12 inches apart. Run the bed or the rows north and south to take advantage of the sun's track from east to west. If rows are used, make a wide row of about 3 feet or more across and plant the corn in a block down the row. That is, rather than have single rows of corn between which you walk to cultivate and water, grow the corn in what amounts to a long bed. Make room to cultivate and water on each side of the bed. This method insures better corn pollination than single rows, and keeps the soil from compacting under foot. About 20 feet of this kind of row will be more than adequate for a popcorn-crazy family of four.

If you want a more modest crop, or you're growing popcorn as a project for your children, make a shorter bed or plant four to five hills, far enough apart to walk around. Plant six to eight seeds per hill, and thin out to four plants when the corn is up. The size of the hills should be about 2 feet square.

Popcorn requires a lot of water. Test the moisture periodically by digging down with your finger. The soil should be just damp. Never let the soil dry out so much that the leaves of the popcorn begin to droop. But try not to overwater, either. Overwatering can be harmful if water stands around the roots of the plant.

Popcorn takes longer to mature than sweet corn—about 90 to 100 days, although it's hard to be precise. Variety, temperature, hours of sunlight and amount of moisture all influence the length of time from planting to harvest. In any event, the long growing season means that the nutrients you added to the soil in the fall or winter are going to be depleted by midsummer. To keep your popcorn in prime condition, side-dress with compost six weeks after planting, and again at the time the silks appear. You might, in rich soil, get away without this extra feeding, but you'll have taken out insurance that you will get good-sized, well-filled-out ears if you make these extra fertilizings through the summer.

Some gardeners find that their popcorn is seldom bothered by pests like sweet corn is. But there are precautions you can take to guard against insect invasion. The European corn borer, that most troublesome pest which lays its eggs in cornstalks, can be kept away by removing the stalks from the garden or plowing them under immediately after harvest. Corn earworm, the destructive little demon that attacks the buds, husks, and silks of young corn, and the kernels of older corn, can be controlled by putting a few drops of mineral oil into the silks at the tip of each ear. But wait until the silks have wilted and are turning brown before using the oil. Earlier application interferes with pollination and can result in poorly filled ears.

When the corn is mature, let the ears dry on the stalk until the stalk itself is dried out. Strip off the ears, husk, and store them in a cool, dry place. When you get the time, it's better to strip the kernels from the ears and store them in a sealed container. This practice will serve both to keep insects off the ears, and make for more compact storage. Holding the ear firmly in both hands and using a twisting motion usually accomplishes the stripping neatly. Some types of popcorn have sharp ends that demand you use gloves when stripping ears by this method.

Keep the kernels in containers for several days to allow the

• Popcorn is the ideal plant to get the kids involved in gardening. To give them a personal interest in growing it, turn the popcorn care and feeding over to them. From the time they plant the seeds in the ground to the time they have their first popcorn party, this project is all theirs. Few other vegetables can be matched in appeal and easiness of growing, and what a dramatic conclusion to all those months of waiting. You could say it all ends with a bang.

moisture held inside them to even up. Use extra care in handling the popcorn to assure yourself that the kernels are drying out evenly. Otherwise the corn will pop unevenly—some kernels popping quickly and some not popping at all—and it will be troublesome to prepare. For long-term storage, you'll need to employ the same precautions necessary for storing other grains, to prevent insect eggs from hatching out and ruining your popcorn. Storing in jars in the refrigerator or vacuum packing are two good ways to store popcorn for long periods of time.

Popcorn has always held a great fascination for kids. They have fun growing it, and they like eating it even more. Sooner or later, they all ask the perennial question—"what makes popcorn pop?" Here's how to answer them: The small, hard kernels, while perfectly dry on the outside, trap moisture inside themselves. When the kernels are heated, the moisture in the cells expands until it creates so much pressure that the kernels explode and are actually turned inside out.

How to make reluctant popcorn pop better is a matter of great debate. Some gardeners recommend sprinkling a little water over it in the jar, replacing the cap, and shaking the jar so the kernels absorb the moisture evenly. After a few hours their popcorn is poppable. Other gardeners find that the popcorn won't pop, not because it's too dry, but on the contrary, because it's too wet. Their solution to the problem: spreading the popcorn in shallow pans and drying it slowly for a couple of hours in an oven set at lowest temperature. Still other folks claim that the corn should never be dried under artificial heat or the "poppability" and the flavor will be ruined. Instead, hang the ears to dry in an unheated shed. If you have problems getting your corn to pop evenly, experiment with the above techniques until you find the one that works for you.

We're all familiar with popcorn as a buttered-and-salted snack. But did you know how good a snack it really is? It's low in calories, and contains protein, food fiber and B vitamins as well as calcium, iron and phosphorus. For a change from the butter-and-salt routine, try sprinkling your hot popcorn with grated cheese, garlic salt, your favorite herbs, or brewer's yeast.

You can use popcorn for other things besides snack food,

Raising Popcorn in the Greenhouse

A friend sent us a couple of ears of what he called "Strawberry Popcorn"—the strangest thing we'd seen in some time. The ears resembled huge strawberries, mahogany-colored and full of small kernels. The stalks grew about 4 feet high, and each stalk had three or four ears. Even the kernels popped beautifully.

We sowed the seed in pots and directly in the bench, spacing each 2 inches apart, using a soil mixture of one part each of sand, peat and loam. In the greenhouse, growth is fast! The stalks started showing ears before we knew it. We let the ears ripen and found out that they can be used for centerpieces, door swags, and fall or winter arrangements.

—Doc and Katy Abraham
Naples, New York

too. One Michigan gardener found that it makes great corn meal. Here's what he says about popcorn meal:

> It is really different and has a mild flavor. This year I raised a good quantity of SOUTH AMERICAN GIANT, and it makes the most beautiful, deep yellow meal. It is wonderful in Johnny cake and mush. I'll never buy corn meal again as long as I can grow and grind popcorn.

You can also use popped corn as the base of an unusual custardlike pudding by soaking it in milk overnight, mixing in beaten eggs, salt, sugar, and vanilla, and baking it until the custard sets. An Arkansas gardener whose hobby is making corn husk dolls says that popcorn husks work a lot better than the husks of sweet corn for making the dolls. Strawberry popcorn makes a colorful addition to fall and winter dried arrangements.

Popcorn seed can be purchased from many commercial houses, including D.V. Burrell Seed Growers Co.; William Dam Seeds; Exotica Seed Co.; Farmer Seed and Nursery Co.; Henry Field Seed and Nursery Co.; Grace's Gardens; Gurney Seed and Nursery Co.; Joseph Harris Co., Inc.; H. G. Hastings Co.; J.L. Hudson, Seedsman; Johnnny's Selected Seeds; J.W. Jung Seed Co.; Earl May Seed and Nursery Co.; Nichols Garden Nursery, L.L. Olds Seed Co.; Geo. W. Park Seed Co.; and Stokes Seeds, Inc. W. Atlee Burpee Co. and R. H. Shumway Seedsman carry especially nice selections of popcorn.

Sweet Potato

Ipomoea batatas

NATIVE to South America, the sweet potato was served to Columbus by the people of the West Indies on his fourth voyage to the New World. Sampling four varieties at a feast, he later enjoyed sweet potato bread worthy of mention to the folks back home. Ethnobotanists say that these fleshy tubers flourished in Peru during the earliest stages of agriculture, and were carried across the Pacific to Polynesia and New Zealand. Much later Spanish explorers took them from the Caribbean across the Atlantic to Spain. The heat- and moisture-loving plant spread slowly through Europe, but it didn't do too well even in the south. Toward the north it was no match for the Irish potato— which is *not,* by the way, a close botanical relative.

Finding a more congenial home in Africa and in the Philippines, India, China, and other tropical and subtropical parts of the Far East, today the sweet potato is the second most important crop in Japan. It's also doing nicely, thank you, in Australia and southern Russia. Here in the United States, sweet potatoes have been a popular southern crop since the mid-1700s, and they are grown commercially as far north as New Jersey. Tasty firm- or soft-fleshed varieties are also being grown successfully by home gardeners, even though this vegetable is vulnerable to root and stem diseases in the field and in storage.

As a relative of the morning glory, the sweet potato sends   *309* forth creeping slender-stemmed vines up to 15 feet long. In the tropics, the plant flowers profusely, producing white or pale violet blooms that set seed. In cooler areas, however, plants don't often flower and they set seed even more rarely. Unlike the yam, which belongs to a different genus and family, the sweet potato forms its tubers as swellings on its roots. High in food energy, these mealy tubers are white or yellow in varieties grown in the United States and range in shape from elongated to nearly round, depending on variety and growing conditions.

To give high yields of short, chunky potatoes rich in carotene, this tropical 'tater needs at least four—preferably five—frost-free months, and warm nights as well as hot days. With early planting so essential to large high-quality tubers, gardeners have had to devise ways of stretching their growing seasons. Indeed, an unusually early start is necessary, for sweet potatoes can't be grown from cut-up tubers as Irish potatoes are: they must be started from sprouts off seed tubers or from vine cuttings from growing plants.

There are varieties aplenty of sweet potatoes grown in this country, just about all of them the dark yellow or reddish orange kinds offering generous amounts of vitamin A. The soft-fleshed, sugary, yam-like varieties more favored in the South include the PUERTO RICO, which has been found to store well. The BUNCH PUERTO RICO is a compact version that does well in the greenhouse. Other southern types are the old variety NANCY HALL (resistant to soft rot), the deep orange mutant NANCY GOLD, HEART-O-GOLD (resistant to root-knot nematode), and the popular new stem-rot resistant ALLGOLD, which also resists the viral disease called internal cork. Among the mealy and starchy "dry" strains preferred in the North are the oldies BIG STEM JERSEY and YELLOW (or LITTLE STEM) JERSEY, which resists root-knot nematode. Newer Jersey-style varieties include MARYLAND GOLDEN, ORANGE LITTLE STEM (also called JERSEY ORANGE), and ROLS, which is a red-skinned variant of Orange Little Stem.

You can grow your crop from mail-order sprouts or from sprouts or stem cuttings you start yourself. Make sure, though, that your seed potato hasn't been treated with a hormone to prevent sprouting in storage. To get vine cuttings, plant whole seed potatoes under 2 inches of light soil or sand in a propagation bed when it's about 70°F. or in a cold frame, hotbed, or greenhouse about two or three months before the last expected frost date. (As a rule of thumb, a temperature of 80°F. in the bed at planting and 70° to 80°F. thereafter will result in slips ready for planting out in about six weeks.) Well aware that warmth means quick sprouting, Japanese growers in areas with long growing seasons use the heat of mini-compost piles to start their seed potatoes. Filling a hole several feet deep with alternating layers of manure, weeds, and straw, they nestle the seed potatoes into the top layer of straw. Another way of getting stem cuttings is to start sprouting whole potatoes in water in a widemouthed jar on a windowsill a few months before you plan to set potato plants out. In all these methods you can break off the 10-inch-long tips of the vines as they develop and transplant them to the garden. By starting your seed potatoes quite early, you can get several slips from each of the several sprouts growing from one potato.

If you get started in a greenhouse or hotbed a mere five or six weeks before planting-out time, you can pull the sprouts from the seed tubers as soon as they have two leaves, wrap the lower part in sphagnum moss, then set them back in the hotbed so the roots that grow out from the main stem become embedded in the moss. (This should be done at least three weeks before it's time to set them out.) If hardened off when the temperature climbs above 50°F. and kept free of weeds when planted out, the sphagnum moss-wrapped shoots are said to get about a three-week jump on the season, outyielding the unwrapped shoots significantly.

You should start preparing an outdoor bed for your vine cuttings or shoots as soon in spring as you can work the soil. The ideal setting is a well-drained sandy loam or loamy sand with a clay or clay-loam subsoil. (Heavy clay soil yields misshapen potatoes, while soil that's too light brings about tubers that are long and skinny.) Your tuber soil should be slightly to moderately acid (pH 5.2 to 6.7), and if it's around pH 5, you should add lime to raise the pH slightly. If you're coping with poor, sandy

• I start slips two months before our last frost. In two short furrows, I lay seed potatoes end to end, and cover 4 inches deep. If frost threatens, I pile hay over the potato bed.

Two weeks after our last frost, I open a trench 4 inches deep, spread rabbit manure and wood ash, cover, and mound soil into a 4-inch-high ridge. Then I twist slips from the mother plant, each slip having at least two leaf nodules. I break long vines into several slips. Roots may already be forming at the nodules. . .

We harvest as many sweet potatoes from clayey soil as from sandy locations. Digging is harder in clay; but rodent damage is worse in sandy soil.

—*Nancy Pierson Farris*
Estill, South Carolina

soil, you'll want to add manure and potash, but go lightly if you've got silt or silt-loam soil since researchers have found that too much nitrogen added to already rich soil can reduce yield, giving you lots of vine and long, thin potatoes to match. Adding potash, by the way, seems to help make shorter, chunkier potatoes only if the soil is deficient in it.

Make the sweet potato rows 18 inches wide and about 3½ feet apart, running a 6-inch-deep trench down the center of each row. Fill this trench loosely with compost or soil fertilized as needed, then use regular garden soil to build a high ridge over the fertilized trench. Good yields are linked to ridges of 12 inches, which keep plant roots in loose soil, making possible big, well-shaped potatoes. But if rainfall isn't heavy in your area and you have good drainage, you can make lower ridges or even plant on the level. Keeping the tops of the ridges wide and flat so they don't dry out quickly, let them settle in a while and plant early vegetables like peas and lettuce in the space between the tuber rows. (These will mature before the potato vines begin to creep.)

About 10 days after your area's last frost date, you can plant your 'taters. Using a hoe handle, poke holes every 18 inches in the ridges, making them as deep as you can and still leave at least one leaf of the slips above ground. Add a little side-dressing of fertilizer to each, be sure to firm the soil around each slip to keep the moisture in, then water well. You should cultivate lightly around young vines during the first few weeks to keep weeds down. When the vines begin to crawl, remove whatever vegetables remain in the aisles and mulch the whole area with hay or straw. Then give it a good soaking. As the season rolls on, you might lift the longer vines once in a while to keep them from rooting at the joints.

One of the biggest problems sweet potato growers can face is rot due to overwatering or poor drainage. There are also, alas, a host of diseases and insects that adore sweet potatoes. Some of these are so contagious that they necessitate scrupulously sanitary growing, storage, and shipping practices. You can protect your crop when it's in the garden by choosing disease-resistant varieties, practicing four-year rotation, making sure

• One gardener, D.V. Davis, has found that frost is actually good for his sweet potatoes:

"When frost comes I pay no attention to it. A week or two later, when the ground has taken all the frost out of my sweet potatoes, I dig them and throw them into baskets and set them in the basement near the furnace, then eat sweet potatoes all winter and the next summer till the new crop comes. I used to dig them early, try all schemes to keep them, but throw most of them out rotten before spring. However I learned you don't have to baby sweet potatoes if you don't dig them too early."

your seed tubers or sprouts are disease-free, and using sand or soil free of nematodes, weevil larvae, etc. Also be sure to handle the plants as little as possible in transplanting and cultivating so there are no wounds to provide a point of entry for disease spores.

Growing sweet potatoes in a greenhouse is tempting in places with short growing seasons—and it's not at all hard to do. They'll do well in large pots or tubs, trained up trellises or wires or strings to the roof bars. To grow them in a bench, place a bottomless box or boxes over the area where you want them to grow, then fill it with soil. Just make sure to leave about 10 square inches per plant. If the pesty white fly should show up, spray the undersides of leaves with a nonphosphorus detergent, using 1 teaspoon to 2 quarts of water.

According to the experts, you can improve both the vitamin A content and the yield of your tubers if you hold off on harvesting. Do, however, dig them up when frost cuts down the vines, for decay on dead vines and rapidly chilling soil can cause potatoes to deteriorate. Since several of the most damaging sweet potato diseases can set in or grow worse during storage, it's important to follow some time-tested procedures after harvesting. Ideally you should cure your crop for four days at 85°F. and at 85 percent humidity. If either the temperature or the humidity is lower, increase the curing time accordingly. Some people cure their potatoes in open shade for 10 days. Northerners might prefer to set them indoors near a hot water heater. This period of heat and humidity hastens the drying of any cut or broken places and reduces shrinkage due to water loss. After curing, the tubers keep best when stored at 55°F. at a moderate humidity up to 75 percent. One gardener reports prolonged storage is possible when sweet potatoes are wrapped individually in newspapers and stored in a barrel. You might also try slicing some potatoes thinly, then drying them in the sun for a few days as the Japanese do—or in a low oven. This method helps preserve sweet potatoes longer and provides surprisingly tasty "chips" for snacking.

During curing and storage some of the carbohydrate in sweet potatoes changes from starch to sugar, improving their flavor with age. And this satisfying taste is accompanied by a whopping 15,000 international units of vitamin A per good-sized raw potato. Containing practically no fat, the sweet potato also offers worthwhile amounts of magnesium, calcium, and iron, and about half as much vitamin C as an orange.

"Sweets" are superb when mashed, baked, boiled, or when they're sliced and batter-fried as tempura. They also make a rich baked pudding and may be used in place of pumpkin in pies or made into flavorful biscuits, candies, ice cream, and cookies. Pureed or just boiled, sweet potatoes can become part of a tasty soup, and grated raw, they go well with chopped green pepper and cucumber and with finely ground nuts and mayonnaise in an unusual salad.

You can order sweet potato sprouts and slips through the mail from Gurney Seed and Nursery Co.; H.G. Hastings Co.; Lakeland Nurseries; Geo. W. Park Seed Co.; and Thompson and Morgan, Inc. The Steele Plant Company in Gleason, Tennessee (38229), carries several varieties of sweet potatoes and yams.

• When my sweet potatoes threaten to shrivel up (due to my lack of proper storage space), I fill my oven with them, bake them, and store them in plastic bags in the freezer. I remove them as required for a meal, either slice or halve them, and fry in bacon fat or butter till nicely browned. They're delicious!

—*Irene Fuderer*
Spencer, Ohio

Lady Godiva Pumpkin

Cucurbita pepo

THE LADY GODIVA pumpkin is a vegetable as intriguing as its name suggests. Along with its other pumpkin cousins, it has been credited with maintaining prostate health, preventing kidney trouble and hair loss, helping resist colds, and postponing senility. Such powers lie in its seeds, which scientists have shown contain 40 percent high quality protein, 45 percent oil in polyunsaturated fatty acids, and other nutrients including phosphorus, iron, calcium, magnesium, trace minerals, and zinc. A study reported by *Agronomy Journal* in 1975 estimated that just over one cup of Lady Godiva pumpkin seeds would meet the minimum daily protein requirement for the average adult.

Since the beginning of the century, Europeans—particularly Germans, Austrians, and Balkans—have grown this vegetable to take advantage of its properties. But it has only been since 1972 that seeds for the Lady Godiva pumpkin have been available in North America, as a result of a United States Department of Agriculture breeding program.

Unlike most other pumpkins, the Lady Godiva has naked seeds. Although all of the layers of the seed coat are present, they do not thicken and harden in the usual way. The basic difference between a naked-seeded variety and its hull-clothed relatives is the seed wrapping: other pumpkins have tough white casings around the seeds, while the naked-seeded type have only a thin green tissue. These seeds are thus more easily accessible; they can be enjoyed without having to be hulled.

Only limited varieties of naked-seeded pumpkins are available. Lady Godiva, with its round fruit, is the most popular. The STREAKER, also grown primarily for its seeds, has oblong fruit. Some people say that the flesh of these two varieties (which contains 25 percent protein) is very tough and more suitable for livestock than for human consumption. But a newer variety of naked-seeded pumpkin, TRIPLE TREAT, with seeds smaller than the Lady Godiva, has flesh as sweet as the most popular pumpkins grown for pies.

The naked-seeded pumpkin belongs to the *Cucurbita pepo* species which also includes most other pumpkins, acorn squash, and summer squashes such as zucchini, crookneck, and Patty Pan. Like some other members of this family, the Lady Godiva grows on vines 8 to 10 feet long. Each vine produces about a half dozen fruits. Lady Godiva's fruit is pale orange or yellowish with green stripes, has a diameter of about 8 inches and weighs about 6 pounds. The seeds are large, averaging ⅝ of an inch in length. Because the crop sprawls, it requires a 30-by-30-foot garden plot to produce 100 pounds of seeds.

Lady Godiva seeds should be planted at the same time as other pumpkins, once the weather has warmed the soil. To help them germinate, soak the seeds overnight and place them between two layers of cotton until they sprout. Because the plant's roots are very sensitive to injury during transplanting, it is best to plant the seeds directly in the garden, at a depth of ¾ inch.

If you are eager to get a head start on the weather, begin the seedlings in peat pots that can later be placed directly in the soil. Or start the seeds in a well-moistened mixture of sifted topsoil, compost, and sand, feed the growing seedlings a weak fish emulsion solution for three or four weeks, and then transplant carefully. To lessen the shock, do this toward evening or on a cloudy day, soak the seedlings well, place them in holes filled with weak fish emulsion, and pack the seedlings in firmly. Although it is contrary to customary advice, this method, used by one gardener, has been quite successful.

But it is certainly less risky to begin by sowing the seeds directly in the garden. Pumpkins do best in a soil which is slightly acid, light, and rich in organic matter. In watering and fertilizing, keep in mind that the roots spread very widely in the top foot of soil, and be sure to cover this area well. Irrigate Lady Godiva plants every week or 10 days and fertilize regularly with fish emulsion or organic fertilizer for best results. Go easy on the nitrogen, however, or you may get very lush plants and few fruits.

Lady Godiva pumpkins are susceptible to the same diseases and pests as all members of their family. Squash or stink bugs tend to hide under things at night, and can be picked off in the early morning from the undersides of boards laid down near the plant to attract them. Squash borers attack the plant's stem and you'll notice it go limp. Slit the stem carefully to evict the creature and cover the cut with soil to repair. New roots will grow. Young plants may be destroyed by the striped cucumber beetle which carries wilt disease. To keep this pest at bay, try surrounding the plants with screening for protection while young.

As the vines grow, keep an eye out for yellowish spots, brown or shriveled leaves. These are signs of mildew. Remove any affected leaves to prevent the disease from spreading.

It is reassuring, despite these potential problems, that Warren Tilsher of southern California has had few problems with his Lady Godiva pumpkins. "After over 50 years of intensive study and experience in growing all manner of plants," he comments, "we believe that they, like animals and humans, have means of protecting themselves against pests and diseases—*if* they are grown in soil well supplied with organic matter and the necessary minerals. We have seen many cases where plants were attacked by insects, yet somehow recovered without our resorting to the usually recommended insecticides."

For the pulp especially, try harvesting the Lady Godiva pumpkin when not quite ripe. Margaret Bassignani of Norfolk, Massachusetts, does this when the pumpkins are 6 inches across. She finds the flesh even more delicious than summer squash and the texture of raw slices with seeds better than avocado.

Fully ripe pumpkins will be ready for harvest 110 days after planting. When the pumpkins are about 8 inches in diameter you may notice cracks at the attachment of the fruit to the vine. That means they're ready to pick.

The shell of the ripe Lady Godiva pumpkin is very tough—so tough that one gardener, when he has large numbers, splits them with an axe rather than a knife. But the hard shell makes storage practical, in a cool, dry place, through most of the winter. They can then be used as needed.

Not requiring hulling, Lady Godiva seeds can be easily dried. Simply cut the fruit, scoop out the seeds, and rinse well. Alternatively, the fruit can be cut in quarters and the seeds and flesh soaked in water where they will separate. Place the seeds on a tray or screen in a thin layer and put them in a warm place. It takes several days to a week for the seeds to dry thoroughly.

Seeds can be eaten raw, as a healthy snack with a delicious nutty flavor. Try grinding the pumpkin seeds with sunflower and sesame seeds to make them into a peanut butter-like spread. Or add them to granola. Some people claim raw seeds are the most nutritious, and that roasting destroys enzymes, lowers the vitamin content, and affects the health-giving properties. But roasted, the seeds still provide many nutrients and a taste superior to that of many nuts. To roast, place the seeds in a 350°F. oven for 15 to 20 minutes. If you want to salt them, do it before roasting, while the seeds are still moist and the salt will adhere more easily.

Why not try some Lady Godiva pumpkins in your next garden? Remember, it is one of the richest sources of protein you can grow.

Seeds for Lady Godiva are available from W. Atlee Burpee Co., William Dam Seeds, Joseph Harris Co., Inc., and Grace's Gardens; for the Streaker from Stokes Seeds, Inc.; and for Triple Treat from Burpee.

• Hulled pumpkin seeds are selling at over $2 a pound, and at that price it pays any health-minded gardener to grow his own—especially when he can now have them practically ready-to-eat. •

—*Warren Tilsher*
Rosemead, California

Purslane

Portulaca oleracea

PURSLANE is a green vegetable of international standing. Mexicans call it *verdolaga* and harvest it from fields where it grows wild for use in soups and salads. The Chinese use purslane in stir-fry dishes and call it *carti-choy*. Italians call it *portulaca,* Arabs and Filipinos eat it, the French package *pourpier* seeds to grow in home gardens. But to most Americans purslane is just a weed.

Native to India, for centuries the plant has been making its way around the world. It is used where it grows wild—Mexico, Europe, China, the Middle East—but a cultivated form is also enjoyed in England, Holland, and France, and has been for centuries.

There are two varieties of purslane. The wild form is a sprawling plant that grows low to the ground, perhaps 2 inches high. It spreads its reddish green or purple tinted stems about a foot across. Wild purslane leaves are greenish purple. Cultivated purslane, from France, is a more erect plant, with larger leaves of a golden yellowish color. Both varieties have thick fleshy branches and succulent stems which ooze fluid when squeezed. The fat, spoon-shaped leaves are between ½ and 1½ inches long. The plant usually bears tiny bright yellow flowers, but some types have white, red, or orange blossoms instead.

Like other weeds, purslane is very hardy. Its persistence lies

323

in an ability to retain moisture and store energy in its stems, thus enabling the seeds to mature even after the roots have been removed from the ground. This capacity gives purslane a mysterious way of traveling great distances. It spreads so easily that you may think you've contained it in your backyard only to have it come up again in the front lawn. And once established, the joints of purslane stems will produce roots wherever they make contact with the soil.

Purslane will thrive in almost any kind of soil. One gardener, to his dismay, raised healthy plants in a can of sand into which he had thrown some. Although young seedlings are less hardy, once established, these drought-resistant plants need little care.

Purslane does not have to be planted every season, as it will reseed itself in the same spot from year to year. Keeping the crop under control is likely to be your major task in cultivating purslane. The simplest solution is to grow it in containers to keep it from spreading wildly.

There are several ways to begin cultivating purslane. You can probably find plenty growing wild near your garden. Look along roadsides, in old pastures or garden plots, on rocky bluffs, or even in the cracks in the sidewalk. Pick some plants to transplant in your garden. Another way to start purslane is to gather the seeds from a blossoming plant and sow a row in your garden. Plant them in the spring after all danger of frost has passed. If you want to grow the cultivated French variety of purslane, seeds are available from a few suppliers.

For the most tender crop, purslane should be harvested before the plant has developed flowers. The leafiest new growth is the tastiest part to eat. Pick the succulent stem tips regularly for a continuous crop and the plant will keep producing right up until frost. But once flowers have developed, growth will cease and the stems will toughen.

One gardener who has grown purslane says it is best to think of it as a seasonal green since storage is impractical. But another remarks that it freezes very well, like spinach, for winter use. Why not try it both ways?

Purslane has a slightly acid taste due to its very large amount of vitamin C. In fact, it contains more vitamin C than an

• To my alarm my five-year-old grandchild swooped down upon a weed growing out of a crack in the public sidewalk, popped it in his mouth and began to chew. "Purse-ulane!" he smiled as he ate it as though it were a sweet. I let him get away with eating off the street because the damage, if any, had already been done. I did attempt to convince him that "food" found on the sidewalk was not to be trusted. After all, I was responsible for having taught him that purslane growing wild in the garden was good to eat. While picking vegetables we had often sat in the shadow of the pole beans to nibble purslane and discuss world affairs.

—John Meeker
Gilroy, California

equivalent amount of orange juice. The cultivated form has a somewhat milder flavor.

Purslane's pleasant nip adds spice to a meal. It can be enjoyed raw or cooked. In salads, purslane leaves are delicious, and the yellow-leafed French variety also adds a spark of color. It makes a tangy dip when 1 cup of chopped purslane leaves are combined with a half cup of yogurt, some chopped onion, salt, and a dash of cayenne pepper. The whole plant can be used if purslane is to be cooked as a green. Boil for only 5 minutes and serve with butter and a touch of lemon, or serve scalloped, with a cream sauce and bread crumbs. Or, like the Chinese, try purslane in stir-fry dishes. Fry it quickly with bean sprouts, for example, and a little sesame oil for seasoning. The fat stems of purslane are particularly good pickled to be enjoyed as a condiment. Purslane's mucilaginous texture gives it thickening power and makes it a welcome addition to many soups. You can use it instead of okra in gumbos and creole dishes.

From experience, one gardener/hostess recommends not telling guests what they are eating until after they have exclaimed over purslane's rich, almost gelatinous flavor. Then she tells them they are eating Indian cress, one of purslane's common names.

Purslane leaves are also used medicinally. They contain tannin, phosphates, urea, and various minerals with a large amount of magnesium. Purslane has been used to counteract inflammation and destroy bacteria in bacillary dysentery, diarrhea, hemorrhoids, and enterorrhagia.

As hardy as it is, and needing almost no care, purslane is also grown along superhighways to prevent erosion on the roadsides.

Seeds for the French type of purslane are available from William Dam Seeds and Le Jardin du Gourmet. The common type can be had from Casa Yerba and Nichols Garden Nursery. J.L. Hudson, Seedsman has a type they call winter purslane.

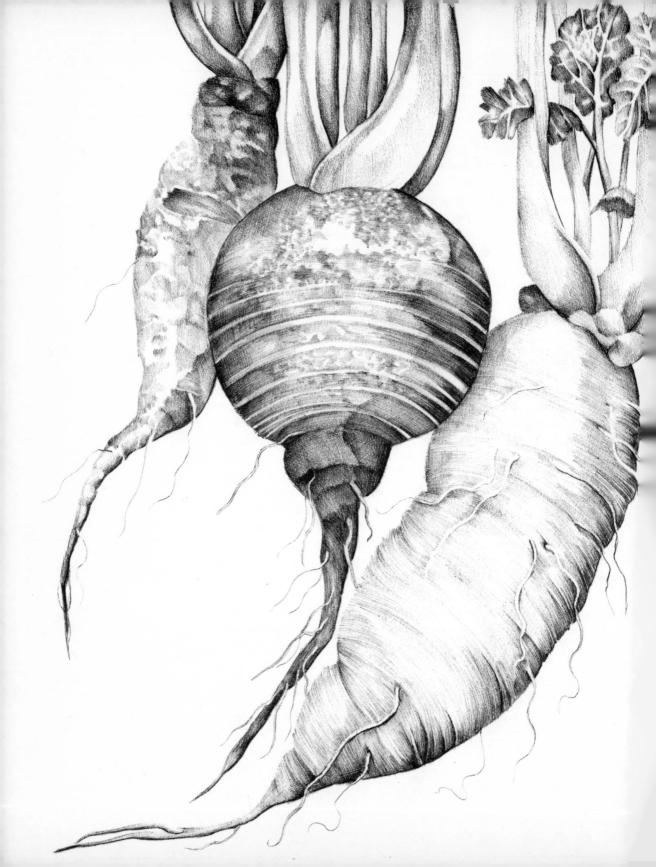

Winter Radishes

Raphanus sativum

"THE first time I saw the big, bold, handsome lobed and toothed leaves of mature giant radishes develop in my garden, my instinct was to pull them all up to use or freeze right away. That sort of harvesting is not necessary, however, as I soon learned, because the big tasty radishes stay edible and delicious for several weeks after maturing. If planted late enough in the season, their 60 to 80 days of growth will bring them to maturity just in time to lift and store over the winter in moist sand."

Ease of harvest and storage, as described above by Catharine O. Foster of Vermont, is one of the best reasons to grow winter radishes. The other is flavor. These big, crisp radishes range in taste from mild to pungent, and can be prepared in all the ways you cook turnips, in addition to having their own special uses.

Winter radishes are grown in late summer and harvested in fall. Why, then, are they called "winter"? Because after harvest, they can be stored for most of the winter without becoming hollow in the middle as spring and summer radishes do when you try to store them raw for long periods of time.

There are many shapes and sizes of winter radishes from which to choose. CHINESE WHITE CELESTIAL comes highly recommended for its subtle flavor. It is round and pure white, growing about 3 inches in diameter and 6 to 8 inches long. Maturity is reached in 60 days. TAKINASHI also earns high marks. Like the White Celestial, Takinashi is white-fleshed and matures in 60 days. Both these radishes are quite mild and can be pulled before maturity if you like. They are always good to eat raw, even after winter storage. A similarly delicious variety is MIYASHIGE, which grows about 15 inches long and 2 inches wide. NERIMA LONG NECK is another tasty white radish which, along with Miyashige, is especially good for slicing and stir-frying. All these radishes are sometimes referred to as *daikons,* a Japanese generic term for radishes of this sort. *Daikons* are always white-fleshed and mild-flavored.

Without a doubt the most remarkable-looking winter radish is the SAKURAJIMA. Said to be the largest radish in the world, it looks like a giant turnip, and under ideal conditions it can weigh up to 50 pounds. In the United States it is often possible to grow 10- to 20-pound specimens. In the garden, Sakurajimas need as much growing space as tomato plants. They mature in a rather amazing 70 days, and should not be eaten raw at any stage of growth. One gardener who has tried it passes on the following advice: ". . . when I decided to sample one of the smaller, less developed ones raw, I found out that it's not such a good idea. Take my warning and don't. They bite your tongue so sharply that you can barely taste the radish flavor. Cook them instead."

Another turnip-shaped radish is the SHOGOIN ROUND GIANT. Despite its formidable name, this one is smaller than Sakurajima, weighing in at 4 to 5 pounds. It is not quite as pungently flavored, but like Sakurajima, it is best eaten cooked. The Shogoin matures in 65 days.

Two other pungent inhabitants of the radish patch are the LONG BLACK SPANISH and ROUND BLACK SPANISH varieties. Both have a dark brown to black skin and crisp, creamy white flesh. They are about medium in size compared to other winter radishes, and mature in 60 days.

For adding color to salads, nothing can beat the delicious, lavender colored RADISH DE GOURNAY, a long type that matures in 65 days; or the bright, reddish pink CHINA ROSE, which attains a length of about 6 inches and matures in 55 days.

Radishes are no newcomers to the garden—they have been cultivated since ancient times. They are believed to have originated in China, but have turned up all over the world at one time or another. The early Greeks were one people that enjoyed radishes and indeed, appreciated them so much that they cast small models of the vegetable in gold. Beets, on the other hand, rated only silver statues, and the lowly turnip was modeled in lead.

Winter radishes apparently made their way to Europe about 300 years before the small spring and summer types that are now so popular, being reported in Germany as early as the thirteenth century. Later, winter radishes were among the first crops brought to the New World by the Spanish conquistadores.

All winter radishes are cultivated in basically the same way. They do best if planted in very loose, deep, soft soil so that their roots can grow without crowding or warping. Keep them well watered so that they grow quickly and steadily. Setbacks tend to make them tough and woody. Avoid high-nitrogen fertilizers that will promote top growth; instead, enrich your radish bed with phosphate- and potash-containing materials to encourage good rooting.

Planting time is important, as hot weather makes the roots woody, and causes the plants to bolt to seed. Young radish plants enjoy warm weather, but the roots need to mature when it's cool. In northern areas, this means planting in mid to late July to obtain a crop ready for harvest around the time the first frosts set in. In regions where winters are mild, an early spring crop can be had by planting radishes two months before the last expected frost. Or you can plant in fall for a winter crop. It may take a bit of experimenting to determine the best planting date for your area. One southern gardener planted his winter radishes too late in spring and found that they grew 12-inch-long leaves but only finger-sized roots after a long time in the ground. It might be a good idea to stagger your plantings the first year and see which produces the best results.

When planting winter radishes, sow seed 3 to 4 inches apart, in rows about 18 inches apart for larger round types. Remember that Sakurajimas need even more room, and give them a couple of feet. Thinning is imperative for a successful

Winter Radishes, Southern Style

Here's a different twist to preparing winter radishes. First, cut the leaves from the stems, discarding any discolored leaves and most of the stem. Wash the good leaves and slice them into half-inch strips. Then peel the roots and cut them into chunks. Heat a tablespoon of oil in a heavy saucepan, stir in about a quart of greens, stir to coat, then add the chopped roots and some water. Bring to a boil, cook rapidly until the greens wilt, then lower the heat and simmer until the radishes are fork-tender.

crop. It helps rapid growth and also controls warping. Every few radishes in a crowded, unthinned row will turn out to be nothing but tough, undeveloped strings of fiber—no use at all. But a good, mature, 60-day radish that has grown quickly in loose soil will be tender and juicy.

These radishes are little bothered by pests. About the only thing that may attack is root maggots, and their invasions can be halted with applications of wood ashes.

Winter radishes left in the ground will remain edible for several weeks after they reach maturity, but then the fibers slowly begin to toughen. For this reason, it's best to dig your radishes for winter storage even though they can survive temperatures in the teens. In light, loose soil harvesting radishes presents no problem. But it can become a tricky affair if your soil is heavy. Careful use of the spading fork is probably the best way to avoid breaking the roots and leaving some of your crop in the ground.

Once dug, winter radishes can be stored like other root crops in a root cellar, straw-lined outdoor pit, or in a box of cool, damp sand or sawdust. Some folks prefer to stir-fry the dug, cleaned radishes for a few minutes to stop the enzyme action, then cool and freeze them for future use. In this way, they will be handy for any number of Japanese dishes you may want to prepare, and for an incidental vegetable throughout the winter.

Winter radishes can be used in many ways around the kitchen. As a rule, you can substitute them for turnips in any recipe. The mild *daikon* types can be grated raw into salads or sliced into soups. In Japan and China, radishes are quite often pickled in large tubs of brine, much the same way we pickle cucumbers. And of course, winter radishes are a standard ingredient in stir-fry dishes, combined with meat and/or other vegetables.

Daikon leaves make a delicious green vegetable as well. They are tender at any age, and are said to have a wonderful rich flavor that's better than turnip tops and similar greens. Use them in soups or vegetable dishes as you would any other greens.

• The first time I grew winter radishes, I planted Round Black Spanish. With the flurry of fall clean-up and winter gardening, cutting firewood, etc., I forgot about the winter radishes.

When I pulled them, I noted they had mighty dark, rough skins for turnips, but just cooked them—roots and tops, like I cook turnips.

Then I remembered the Round Black Spanish radishes I'd planted alongside the regular turnips. We've planted winter radishes ever since. They provide a bit of variety.

—*Nancy Pierson Farris*
Estill, South Carolina

Winter radishes are available from most seed companies. Kitazawa Seed Co., Nichols Garden Nursery, and R.H. Shumway Seedsman all carry particularly nice assortments.

Rhubarb

Rheum rhaponticum

RHUBARB isn't generally classified as a fruit, but it serves a similar purpose. The plant is grown for its relatively large, thick leafstalks, or petioles, which are primarily used in sauces and pies. While still not a standard vegetable grown in American gardens, its tart, pinkish stalks are becoming a more and more common sight. In fact, it is estimated that over 200,000 tons of rhubarb is grown in the United States annually. That's a lot of pies.

Rhubarb is a native of the colder portions of Asia, probably Siberia. However, several edible varieties have been found growing wild in the eastern Mediterranean area and Asia Minor. The earliest records of the use of rhubarb date from about 2700 B.C. in China, where the root was often used for medicinal purposes. Reportedly Marco Polo stumbled across this ancient plant while on his famous trek through that country.

The rhubarb wasn't introduced to Europe until the early 1600s and not until 1778 was it definitely recorded as a food plant in England. Massachusetts settlers and other New England gardeners were the first Americans to catch on to rhubarb's wonderful taste in tarts and pies, but even then the plant was not grown extensively until 1822.

The rhubarb plant is an herbaceous perennial. The underground portion consists of large, fleshy, and somewhat woody rhizomes and a fibrous root system. The first leaves grow from the crown. The stem is hollow and has conspicuous nodes and relatively small leaves. The flowering stem, which can grow 4 to 6 feet high, develops numerous small, greenish white flowers on slender pedicils. Under favorable conditions some varieties will produce incredibly large plants with leaf stalks up to 3 feet long and thicker than a broom handle. The leaf blades themselves can grow up to 3 feet across.

If you remember nothing else about rhubarb, remember that this vegetable is one of the few of which the petiole or leafstalk is the only edible part of the plant. The root contains potent substances which can cause violent digestive problems and the leaves contain enough calcium oxalate to cause serious illness and even death if eaten.

Gardeners have several varieties of rhubarb from which to choose. VICTORIA and LINNAEUS are the oldest and probably the best known. They are both large and productive with light green or crimson-streaked leafstalks. The Victoria is the principal variety used for forcing because it produces large yields and good-sized stalks. Some of the newer varieties are becoming increasingly popular mainly because of their attractive and succulent red stalks. Among these are McDONALD, RUBY, VALENTINE, and SUNRISE. The McDonald, developed at McDonald College in Quebec, grows well in most parts of the country. Its brilliant red stalks are perfect for brightening up a sauce or pie and besides being good-looking, the stalks do not have to be peeled.

Rhubarb loves a cool climate. Its crowns and rhizomes are resistant to both cold and dry conditions. The plant thrives best in areas where the crown remains frozen all winter and where the soil remains dry throughout the summer. In regions with mild winters rhubarb will grow during the winter months but will remain dormant during the summer. In colder areas the plants will lie dormant in winter but will resume growing in spring and summer. When the thermometer drops below 26 or 27°F., the vegetative parts of the plant die. The stalks tend to develop their pinkish color when the plant's growing temperature is low, while at high temperatures it is the green color that dominates. In the United States the rhubarb is not well-suited for southern states where the mean temperature is well above 75°F. in summer and 40°F. in winter.

The rhubarb, or pieplant, does best in rich, well-drained soil that has been heavily mixed with organic matter. But the plant isn't finicky. It will also prosper in soils of sand, peat or clay and while it is tolerant of most soils' acidity, it prefers a slightly to moderately acid pH. It is important to prepare the soil properly since rhubarb are heavy feeders; they require large quantities of nutrients for good growth and high yields.

The rhubarb does not come true to seed, so for best results propagation should take place by division of the crowns. Seed houses usually supply crown or root starts, rather than seeds. It is possible to grow rhubarb from seed but most gardeners do not recommend it because the seedlings often bear no resemblance to their parent plants. This method is generally used to develop new varieties.

Setting the plant is either a spring or fall event. Spring is usually preferred in most regions because of the advantages of having increased rainfall. If you have chosen a spring planting, be sure to get the crowns in as soon as old man winter has left; the earlier the better. The secret is to set plants just as they are awakening from their winter dormancy. Otherwise it is probably best to wait until fall or the following spring. Where severe freezes do not occur and the autumn season is long, planting can be done during the fall after the foliage has been killed by frost.

A piece of rhubarb plant for planting in the garden must contain some of the large fleshy root combined with some of the compact underground stem structure and buds from which the leaves arise. As a rule, try not to set plants with new stalks any taller than 6 inches. When very old crowns are used, only the vigorous outer portions should be planted. The clumps of sprouts should be set in a prepared trench no deeper than 2 or 3 inches below the surface. Separate the plants in their rows by 2 to 4 feet, and the rows by 4 to 6 feet. When tall enough rhubarb should be mulched with strawy manure, wheat straw or hay.

• If you wish to force rhubarb in the greenhouse, prepare a planting box by mixing ample amounts of rotted manure or compost with soil. Then dig up the rhubarb roots, place them in the box, and work 3 inches of soil over the roots, between the clumps, and around the rhizomes. Keep the soil moist, but not soggy. The best growing temperature for rhubarb is around 60°F. Higher temperatures mature the crop earlier but the color and quality are not as good as when rhubarb is forced at lower temperatures. On the other hand don't let the temperature drop below 50°F., as such low temperatures will retard the plant's growth. Forced rhubarb is ready to harvest when the stalks are about 18 inches long.

When the plants are young, keep the soil moist and the weeds down. Once established, rhubarb will survive for years, even when choked with weeds and grass.

If you're looking for an instant crop, rhubarb is not for you. The stalks should not be pulled the first year and should be pulled the second year only to let them grow further. It takes approximately three years for the plants to reach a fairly productive stage. Harvesting can then take place about one month after the weather starts to turn warm and the stalks are 1 to 2 feet long, and may continue up to eight or ten weeks. Pull the stalks to harvest them, don't cut. Each plant should reward you with 2 to 3 pounds of tangy rhubarb.

Some gardeners prefer to grow their rhubarb for forcing. After the plants have become large and sturdy in the field, the underground parts are taken up in late winter before growth starts and planted in heated, dimly lit houses. Rhubarb can be forced in the basement, hotbed, cold frame, or in any dark or semidark location where moderate temperatures of about 60°F. can be maintained. Light is not essential or even desired. If you plan on forcing, remember that the roots must go through a low temperature period, freezing or below, before they can be brought indoors. Probably the easiest method of forcing, if you have a portable cold frame, is to place the frame right over the plants in the garden, and cover it with a blanket to keep out frost.

Insects do not often plague the rhubarb. The rhubarb curculco, a kind of beetle, will sometimes cause problems by boring into the stalk, crown or roots. But it is easily controlled by hand-picking the intruders and burning the infested plants. Rhubarb also has a tendency to catch crown or root rot caused by a fungus attacking the base of the stalk. The best prevention for this is to make sure the crowns you set are disease-free, and to avoid planting where the soil has constantly been used for rhubarb over a number of years.

Whatever method you choose to grow the rhubarb, you can enjoy this ancient pieplant in savory sauces to accompany meats or to serve with ice cream, in pies or tarts, either alone or combined with strawberries or raisins, and in other creative ways. It makes a mouth-watering fruit crisp or custard pie, and baked into a sweet bread like banana bread, it is truly marvelous. Harvested in spring, rhubarb is nature's pick-me-up tonic after those long and cold winter months.

If you find yourself unable to keep up with the harvest, you can freeze your excess rhubarb without much trouble. Just discard any woody stalk ends, wash the stalks and chop in 1-inch pieces. Blanch for 1½ minutes and pack. Or pack the unblanched rhubarb and cover it with a sweet syrup before freezing.

Rhubarb is available from Burgess Seed and Plant Co.; W. Atlee Burpee Co.; William Dam Seeds; DeGiorgi Co.; Gurney Seed and Nursery Co.; Charles C. Hart Seed Co.; H.G. Hastings Co.; J.L. Hudson, Seedsman; J.W. Jung Seed Co.; Lakeland Nurseries; Earl May Seed and Nursery Co.; Mellinger's, Inc.; L.L. Olds Seed Co.; Redwood City Seed Co.; Seedway; R.H. Shumway Seedsman; Spring Hill Nurseries; and Thompson and Morgan, Inc.

Rocket

Eruca sativa

SANDRA Kocher of Worcester, Massachusetts, discovered rocket by accident. "A packet of rocket seeds caught my eye one day in a spring display of seeds. Ah, a new vegetable! And one with Mediterranean provenance, judging by the alternate Italian and French names on the seed packet—*rucola* and *roquette*."

Native to the Mediterranean region, rocket grows wild throughout southern Europe. In ancient times, its leaves were eaten raw in salads and enjoyed for both their distinctive taste and supposed aphrodisiac effect. For centuries rocket was cultivated in British gardens and elsewhere before making its way to some of the earliest gardens in New England. Today, rocket is popular with Italians, French, Spanish, Greeks, and Egyptians who use it as a potherb and in salads. But unfortunately, the plant is rarely grown in the United States anymore.

Rocket is a hardy annual related to mustard. It has erect stems which may get as tall 1½ to 2 feet, though they are usually a bit shorter. The leaves of rocket are bright green, long and smooth, and, particularly the lower ones, sharply toothed and indented. Plants grow densely and develop white or yellowish blossoms with deep violet or reddish veins.

Cultivated for its flavorful leaves, rocket is an easy-to-grow early spring green. Its primary need is for cool weather. Sow the seeds as early in spring as the soil can be worked. For a continuous crop, make sowings every few weeks. Plant the seeds about ½ inch deep and cover them with fine soil firmly pressed down over the row. Space rows about 16 inches apart. When the plants are 2 or 3 inches high, thin them to stand one every 6 or 9 inches.

For best results, rocket plants should grow quickly and steadily. There are several ways to assure this: sow them early, while the weather is cool; plant the seeds in a loose, well-composted soil; and be sure the plants are kept moist. Rocket's very strong flavor will be tempered by rapid growth.

The leaves will be ready to harvest six weeks after planting, when the plants are between 8 and 10 inches high. Pick the leaves before hot weather has hastened blossom formation. If flowers do develop, nip them off to prolong plant growth. Continuous harvest of leaves will result in the most prolific crop.

In addition to providing fresh greens early in spring, rocket is an excellent late crop as well. In most areas, it can be sown again early in the fall and is able to withstand light frost. As the weather turns cold, the plants can be protected with cold frames, or lifted and grown indoors in pots. In regions where the winters are mild, rocket may be planted later into autumn.

Since it is quite easy to extend the season, enjoy fresh rocket by picking the leaves when you're ready to use them. When summer or winter weather does strike, you can freeze the rest of your crop for later use, in the same way as you'd freeze spinach or other greens.

The flavor of rocket is quite distinctive—sharp, spicy, pungent. Some folks say it resembles horseradish. To enjoy rocket at its best, serve it the way Italians prefer it, raw in salads. The raw leaves add flavor to any salad and are excellent with tomato dishes, too. For a different approach, rocket can be cooked lightly. Or you might try steaming young rocket leaves with other greens—mustard or turnips, for example—for an unusual combination of flavors. Older rocket leaves, which have become too bitter and stringy to eat by themselves, are good cooked and pureed in a blender to add to soups, or to use in their very own cream soup, with seasonings and some sour cream or yogurt added.

Seeds for rocket may be listed under the names roquette, rucola, arugula, garden rocket, or rocket salad, and are available from W. Atlee Burpee Co.; Casa Yerba; Comstock, Ferre and Co.; William Dam Seeds; DeGiorgi Co.; Gurney Seed and Nursery Co.; Charles C. Hart Seed Co.; H.G. Hastings Co.; J.L. Hudson, Seedsman; Le Jardin du Gourmet; Meadowbrook Herb Garden; Nichols Garden Nursery; Geo. W. Park Seed Co.; and Redwood City Seed Co.

Salsify

Tragopogon porrifolius

IF you've never heard of salsify, it may be because you know it by the more commonly used name of oyster plant, or vegetable oyster. Not surprisingly, the name came about because the flavor of the root, when cooked, reminds many people of oysters. The plant is also known as goatsbeard, because of its clump of narrow, grasslike leaves that grow above ground.

Salsify originated in the eastern Mediterranean area, and has been with us for many, many years. The ancient Greeks and Romans knew of it, and gathered it where it grew wild. Salsify was eaten in France and Germany as early as the thirteenth century, although it wasn't cultivated in gardens there until sometime during the 1500s. The English took to growing salsify both as a vegetable and as an ornamental, and it appeared on the American scene at some point in the 1700s. Early Americans valued the root for its "tonic" properties—salsify was believed to have antibilious, cooling, and diuretic qualities.

Salsify is an easy vegetable to grow for the gardener who has plenty of patience and who knows what to look for when the seedlings first come up. A member of the sunflower family, salsify is a true biennial, producing fleshy roots the first year, and setting purple or rose-colored flowers the next. In England, salsify is sometimes called "John-go-to-bed-at-noon," because of the flowers' odd habit of closing at midday. Salsify

A Special Method for Germination

In this region salsify and scorzonera do not usually germinate under usual conditions. As a consequence, many gardeners here feel they cannot grow these vegetables. The following method allows these crops to be grown regardless of ambient temperatures.

Soak the seeds in cold water for 48 hours. Change the water once. Prepare a unit such as is commonly used for germination tests or pre-sprouting any seed. I use a saucer with several layers of wet paper towels and invert a second saucer over all. The germinating unit is then kept at room temperature during the day and in the re-frigerator at a temperature in the mid to upper thirties during the night. After four or five days the radicles will appear. The seeds then need to be checked twice a day. When most of the radicles are about ½ inch long, plant in the usual way, but *gently*.

The alternating temperatures are essential to success.

—*John A. Shultz*
Chiefland, Florida

flowers are followed by heads of "parachute" seeds that closely resemble the seed heads of dandelion.

The most popular variety of salsify for the home garden is MAMMOTH SANDWICH ISLAND, which matures in 120 days. Its roots are creamy white, and reach 8 or more inches in length, with a diameter of 1 to 2 inches at the crown.

Salsify needs four months to mature in order to produce roots of edible thickness, but otherwise it has many of the same habits and requirements as parsnips and carrots. In areas where the winters are mild, salsify is best planted in June so that it will mature in late October for use as needed all winter long. In northern zones, it should be planted in mid-spring so that it will mature at the onset of cold weather. Since salsify is hardy, in areas where the growing season is quite short the seed may be sown before the last frosts of spring are over, and the plants will continue growing on into the fall, for as long as the weather is warm.

Salsify does best in rich, loose loam that will allow the roots to penetrate easily and expand. Like most other root crops, salsify tends to become deformed in soil that is poorly prepared or rocky. Prepare the soil by loosening it to a depth of at least a foot, working up a clod- and rock-free bed. Salsify likes a nearly neutral or slightly alkaline pH, so add some lime if necessary. The best fertilizer for salsify is one that contains a good percentage of potassium, like seaweed or wood ashes. If such potash-rich materials aren't available where you live, granite dust is a good commercial organic substitute. Do not use fresh manure in the salsify bed under any circumstances. It causes forked and misshapen roots, just as stones do.

It's a good idea to start with fresh seed each year—you can't depend on salsify seed to remain viable for two years. Sow seed sparingly in furrows 1 inch deep, in rows that are 15 inches apart. When the plants are well established, thin them to stand about 4 inches apart in the rows. Because its roots are slender, salsify can be spaced closer together than carrots or parsnips. You may find the seedbed slow to start, but keep it moist, and patience will pay off.

Do be careful the first few times you weed the rows.

Just-up salsify looks surprisingly like grass, and many an un-initiated gardener has been mystified by the utter absence of salsify in the scrupulously weeded salsify bed. After the plants have become established, they look a bit like daylily or daffodil plants, and would thus be equally at home in a vegetable garden or a flower bed or border. Keep your oyster plant well mulched, and water it if the top inch of soil dries out. Salsify is seldom bothered by insect pests.

As the season progresses, each plant develops a slender, hairy taproot which resembles a thin parsnip. The roots develop their best oyster flavor after going through several frosts, so by all means delay harvest until that time if you can. Being biennials, the roots are 100 percent hardy outdoors, and store best right in the rows where they grew, if winters aren't too severe. Cut off the tops of the plants, cover them with a good layer of mulch, and dig the roots whenever you need them throughout the winter. In regions where the ground freezes solid, it may be more convenient to store the roots in a cold root cellar, packed in moist sand. Leave a few inches of tops attached for long-term storage. Salsify will shrivel under ordinary refrigerator storage that doesn't affect parsnips or carrots.

The roots taste best when used shortly after digging. The oyster flavor noticeably decreases with prolonged storage, although the roots are still tasty.

The roots have a multitude of uses. They can be baked, boiled, fried, or served in soups. One thing to remember when cooking salsify is to peel the roots after, not before, cooking. Just scrub them thoroughly and steam until tender, about 45 to 50 minutes. The skins will rub off easily after the roots are cooked. Salsify is one of those vegetables that discolor quickly, so when you're not planning to use it immediately after cooking, put the peeled roots in a bowl of water to which some lemon juice or vinegar has been added, to keep the roots from turning brown.

Salsify can be prepared in much the same way as parsnips. Try the roots sliced lengthwise, basted with butter, garnished with parsley and your favorite spices, and baked in a shallow pan. Try it creamed, braised, or topped with herb butter. Your

Does It Really Taste Like Oysters?

Some oyster plant lovers swear they can't tell the difference between salsify and real oysters when they're prepared this way; others heartily disagree. Find out for yourself by making mock oysters for your family some day.

Cook until tender enough salsify roots for your family, rub off the skins, and mash the roots in a large bowl. Stir in a couple of beaten eggs, salt and pepper to taste, and a sprinkling of your favorite herbs. Form into small cakes, and either dip them in beaten egg, then bread crumbs, or make a fritter batter and coat the patties with that. Fry the cakes in a bit of oil until they're nice and brown, drain, and serve with tartar sauce or lemon wedges. This dish is bound to be a real conversation-maker!

family will probably also enjoy a traditional salsify cream soup. It's made the same way you make oyster stew, but with cubes of cooked salsify instead of oysters. For a real test of whether or not you think salsify tastes like oysters, serve it as "mock oysters" (see "Does It Really Taste Like Oysters?")

If you let some of your salsify remain in the garden throughout the winter, the plants will resume growing in the spring. The young shoots taste rather like asparagus and may be prepared in the same way. They're delicious when served with a seasoned butter or a cream sauce. The new leaves also make a tasty early spring green, either cooked or used raw in salads. Some folks allow the plants to bloom, then use the dried seed heads for indoor arrangements.

Salsify seed is available from many seed houses, including Burgess Seed and Plant Co.; W. Atlee Burpee Co.; D.V. Burrell Seed Growers Co.; Comstock, Ferre and Co.; DeGiorgi Co.; J.A. Demonchaux Co.; Gurney Seed and Nursery Co.; Joseph Harris Co., Inc.; Charles C. Hart Seed Co.; H.G. Hastings Co.; J.W. Jung Seed Co.; Earl May Seed and Nursery Co.; Nichols

Troubleshooting Tips for Salsify

Two problems seem to crop up again and again in the experience of gardeners whose salsify crops are failures. The first is the absence of any crop at all, and the second is roots that are so small that after scraping and peeling off the hairs there's nothing left to eat. Here's some advice that may help the novice salsify grower to avoid these common pitfalls.

If your crop fails to materialize, leaving you with empty beds, it's probably because you or a well-intentioned friend inadvertently weeded it out of existence. When it first comes up, salsify looks a lot like grass, and can be easily mistaken for a weed. If you're planning to grow salsify for the first time this year, plant a few seeds in a pot indoors at the same time you sow the outdoor crop, so you'll know what to look for when the young plants come up in the garden.

If you find at the end of the season that your salsify roots are too small to use, the problem may be that the roots did not have enough time to mature. Salsify needs a full four months for the roots to grow large enough to eat. If your growing season is short, this means planting *before* the last frost in spring. Planting earlier next year should improve the size of your crop. Four months is a long wait, but worth it.

Garden Nursery; L.L. Olds Seed Co.; Seedway; R.H. Shumway Seedsman; Stokes Seeds, Inc.; and Thompson and Morgan, Inc.

Scorzonera

Scorzonera hispanica

SCORZONERA is commonly called "black salsify." Although the black-skinned scorzonera is an entirely different plant from the more familiar white-skinned salsify, the roots have a similar flavor and are used in the same ways. Theoretically, the correct spelling of the name should be scorzanera, from the Italian words meaning "black bark" or black rind, but the incorrect spelling is also found in Italy.

Like salsify, scorzonera originated in the Mediterranean area, and was relatively unknown except to wild-food gatherers until the sixteenth century. During the Middle Ages, scorzonera was one of the most important vegetables in Europe. It was considered a potent tonic, and was used in many countries as a remedy for smallpox. Scorzonera was also considered an antidote to snakebite, and some people claim its name is derived from the Spanish *escorzo,* meaning serpent. Louis XIV grew large quantities of scorzonera in his gardens, and depended on it to relieve the attacks of indigestion he suffered.

The Spanish began to cultivate the wild scorzonera in the eighteenth century, and its use spread to kitchens all over the English-speaking world. The vegetable fell out of favor in the Victorian era, it is said, because fastidious cooks began peeling the "dirty" black skin from the root before cooking. Peeling ruined the flavor, and also put an end to whatever tonic

349

qualities the root may have possessed. Today, scorzonera has found its way back into European gardens, and its excellent flavor makes it deserving of more recognition in this country.

Scorzonera's growing habits are pretty much the same as those of salsify, except that the roots are somewhat thinner and have black or charcoal grey skins. During the second year of cultivation, if the roots haven't been dug, the plants develop attractive, long-stemmed yellow flowers similar to dandelions. Like salsify, scorzonera is fairly easy to grow. It will do best in a loose, sandy loam that's been cultivated to a depth of 12 to 18 inches. Compost or well-rotted manure added to the soil encourages larger, straighter roots. If you treat scorzonera just as you do salsify, your crop should be successful. The roots are hardy, and can be left in the ground all winter.

Many gardeners prefer growing scorzonera to salsify. Some claim the roots grow bigger and straighter for them, and most agree that scorzonera has a more pronounced oyster flavor. Scorzonera is especially valuable in the garden where carrot flies are a problem. Planted among rows of carrots, it will repel maggots that would otherwise burrow into the carrot roots. Mixing some scorzonera seed in with your carrot seed before planting is the easiest way to companion-plant the two.

When the roots are ready to harvest, take care when digging them. They are sometimes rather brittle, and may break easily. If cut or damaged during digging, the roots will "bleed" and lose flavor.

Some varieties of scorzonera tend to be bitter, and you may be tempted to peel the roots before cooking to remove the bitterness. Don't do it, or the distinctive flavor of the roots will be lost. If you find you have trouble with bitterness, scrub the roots and soak them in water before cooking.

Like Jerusalem artichokes, the roots of the scorzonera contain inulin, and are thus valuable to diabetics, who can use them as a source of carbohydrate. The inulin is responsible for the sweet quality apparent in the flavor of the roots.

To prepare scorzonera, steam or boil it until tender, about 45 minutes. Drain thoroughly, then rub off the skin. Like salsify, scorzonera must be placed in water to which some vinegar or lemon juice has been added to prevent discoloration, if you're not going to use it immediately. Scorzonera is delicious served hot with melted butter or a cream or mushroom sauce. For a delightful combination dish, cook scorzonera with some carrots and onion, and serve piping hot and buttered. Scorzonera also makes an elegant soup, and is good baked or fried as well.

Scorzonera seed is available from William Dam Seeds; DeGiorgi Co.; J.A. Demonchaux Co.; J.L. Hudson, Seedsman; Johnny's Selected Seeds; and Thompson and Morgan, Inc.

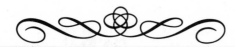

Scorzonera Cream Soup for 6

Cook 6 or 8 large scorzonera roots in water until not quite tender. Peel and chop them into ½-inch pieces. Brown in oil, then add some chopped onion or celery and saute briefly. Stir in a few tablespoons of flour to thicken the broth, then gradually add 6 cups of milk. Season with salt and pepper to taste, then simmer 15 minutes, or until the scorzonera is tender. Serve hot, sprinkled with chopped chives or parsley.

Sea Kale

Crambe maritima

THIS vegetable grows wild along the cliffs and beaches of the English, Continental, and Irish coasts. It is quite a large plant with broad, toothed, bluish green leaves and white flowers borne on a 2-foot-long stalk. Country people have recognized its young shoots and leaves as a delicate and choice food since earliest times, but not until the late eighteenth century was it given a place in the garden. At that time, it was known as "sea colewort" and enjoyed a modest popularity in England and, later, in America. Yet, although it is an easy-to-grow perennial and a delicious food, sea kale has never been as widely cultivated as its relatives the cabbage, cress and turnip. Even in its native countries, it remains something of a luxury item, best known as a forced, midwinter vegetable.

The plants can be propagated by seed or root cuttings.  When seeds are used, they are sown indoors or out, 1 inch deep in a rich, well-prepared seedbed. After the seedlings have become established, they are thinned to stand 5 or 6 inches apart in the row. The following spring, transplant the sea kale to a permanent bed; in two or more years, the plants reach maturity and young leaves and shoots can be gathered for eating. The bed should produce good crops for six or more years.

The quickest way to produce a crop is through root cuttings, but these are difficult to obtain unless you know someone with a sea kale bed. The best cuttings are about ½ inch in diameter and 4 to 6 inches long. These are the prepared straight, side roots or "thongs" which grow out from the main sea kale root. The thinnest end is cut on a slant and the thick top end is cut level. Usually thongs are taken in fall, tied in bundles, and stored in damp sand until planting time. Other growers prefer to make the cuttings in early spring for immediate planting in the garden bed.

Cuttings can be planted in spring for a small harvest the following spring, or in late autumn for harvest the next year. Spring plantings should be made as soon as the ground can be worked, in early March or late February. Plant the cuttings in a permanent bed, 1 inch deep, at 2- to 3-foot intervals. Space rows 3 feet apart. Some gardeners recommend growing thongs in a protected propagating bed until early September, then transplanting to the permanent location.

Autumn plantings should be a bit deeper, and mulched with a thick layer of clean, dry leaves, hay, or straw, and a top layer of earth. Make sure the bed is well covered and the roots protected from frost. To get good shoots in the spring, plant the cuttings during a warm dry spell.

Sea kale prefers a rich, light loam, deeply prepared with plenty of compost and well-rotted manure. To sandy soils, add about a cup of bone meal per plant or fertilize well with fish meal or poultry manure. Seaweed seems to be this plant's favorite food. Direct, full sunlight is also very important for good growth.

Sea kale is harvested in spring, when the shoots begin to develop, and the topsoil begins to crack. The shoots must be blanched to be tender, and in Europe a special sea kale pot is used for this purpose. When you notice cracks in the soil of your sea kale bed, cover the plants with inverted flower pots (with the holes plugged) or boxes, or mound dirt over them. In this way, the petioles are blanched and develop with a tiny leaf at the tip. When the shoots are 6 to 9 inches long, snap them off near ground level and prepare like asparagus. Harvesting can continue until leaves begin to appear. Remove the blanching devices after all danger of frost has passed and leave the plants undisturbed. They will grow to a height of 3 feet. In midsummer the plants will produce attractive flowers, which should be removed. This seldom presents a problem, because the flowers are a nice addition to indoor arrangements. If you live where winters are especially severe, in late fall cut the plants to a few inches above ground level and mulch heavily (as you would for an autumn planting) to protect the roots from frost.

Sea kale can also be forced indoors during the winter. For winter use, dig the strongest crowns in autumn and plant them 2 inches deep in pots or boxes of rich sandy loam. Cover them with another pot or box and place in the dark in a warm greenhouse, windowbox, or hotbed. Water well and harvest the blanched shoots in five or six weeks.

Once a bed has been established, it will continue to produce shoots for autumn and spring cutting for several years. Each year, the bed should be well fertilized with compost and manure. At the close of the season, clean up any dead leaves and debris before covering the plant with a protective mulch.

Sea kale shoots have a delicate, nutty, slightly bitter flavor. They are delicious when eaten raw with cheese or in salads, or when prepared like asparagus. You can also use any cardoon recipe for sea kale. Traditionally, the blanched leaf stalks are tied in bundles and cooked in salted water for about 20 minutes, then served hot with melted butter. Like asparagus, it is served in Europe as a separate course. Available when few garden-fresh vegetables can be had, sea kale is a welcome addition to the menu.

Root cuttings are best obtained from fellow gardeners or growers. Seeds are available from two suppliers—J.L. Hudson, Seedsman; and Thompson and Morgan, Inc.

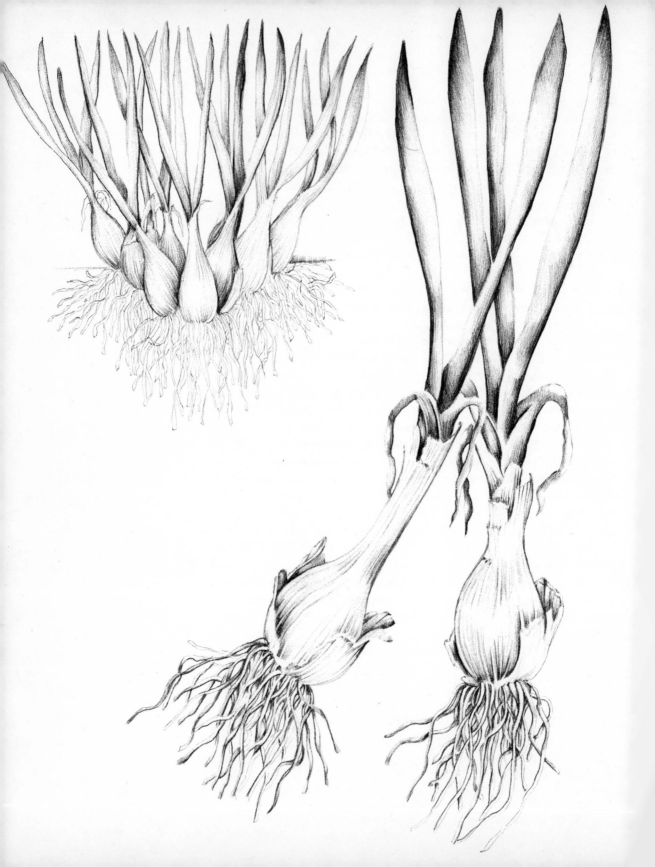

Shallot

Allium cepa, Ageratum group

THE shallot is believed to have originated in western Asia, the name being a corruption of Ascalon, a city of the Philistines. The shallots which grew there were famous in early Christian times and are mentioned in Pliny's writings. They were brought to Britain by the returning Crusaders.

For many years, the shallot was considered a distinct species (*Allium ascolonicum*), but recently taxonomists have re-classified it as a member of the ageratum or lily group within *Allium cepa.* Members of this group are characterized by bulbs which multiply freely through several lateral cloves. Thus, shallots look more like garlic than onions.

Of the several varieties that are grown throughout the world, two are very popular among American gardeners. The Welsh type has a tight-fitting, light brown to tan skin, is a prolific grower and keeps very well. The French-Italian type has a brownish red skin, a stronger flavor, and an aroma that resembles garlic.

Many other varieties are available to gardeners who seek shallots with particular flavors, certain shapes, or resistance to various diseases. Southern growers often select the LOUISIANA PEARL because of its tolerance to pink root. The JERSEY shallot is grown for its unusual shape, the bulb being compressed and sometimes larger in diameter than length. All types are perennial, grown both for the young plants which are used in the same way as green onions, and for the dry bulbs.

Since shallots require 120 days of growing time, they must be planted very early in the spring. If the growing season in your area is extremely short or you would like to get a head start, the cloves can be started indoors in shallow flats filled with a mixture of peat moss and coarse sand. After top growth has appeared and an extensive root system has developed, transplant the shallots to a permanent garden bed. In most regions, the cloves can be set outside as soon as the soil can be worked. There is no need to fear light frosts for, like onions, shallots are quite hardy and resistant to cold.

Choose an open site that is deeply worked and well-drained. A light soil high in phosphorous and potassium but not overly rich in nitrogen is best for optimum bulb development. When planting, bury the clove to half its depth, then firm the soil around it making sure the top remains uncovered. Place the cloves at 4- to 6-inch intervals, in rows 9 inches apart. After five or six shoots have developed, mulch with straw, peat moss, leaves, or dry, unfinished compost. Too heavy a mulch will encourage onion maggot and root rot, but no mulch at all will necessitate frequent waterings and cultivation.

Within a week of planting, the first green shoot appears. Three weeks later, the first bulb division occurs and in three to four months, the tops wither, turn brown, and drop off completely. The drying out, which signals the full development of the shallots and the end of the growth cycle, can be accelerated by withholding water or simply by pressing down the tops after they have begun to lose their strength. Each shallot bulb now consists of five to ten cloves and can be harvested at any time. If dug when the tops are still slightly green and pliable, shallots can be braided for hanging; otherwise the tops are cut off within 2 inches of the bulb. One or two days of sun drying are required before storing on trays, hung from the rafters or in net bags, in an airy, cool place. The shallots will keep until the following May or June.

If grown in a well-drained, sunny location, shallots should not suffer from pest or disease damage. Good weed control and

• In the spring, I prepare 5-foot beds (usually about a foot high for extra drainage) that are well fed, for the shallots respond accordingly. My four rows in each bed are 8 inches apart both ways and I've made a weeding tool that lets one cultivate both ways, very easily.

The shallot sets that you advertise a hundred to a 12-ounce box, I grind up and use for fertilizer. It makes an excellent food for roses and stinging nettle.

—*Gardener's letter to*
Le Jardin du Gourmet
West Danville, Vermont

liberal waterings will take care of most problems.

Needless to say, shallots are a must in many dishes that call for a flavor subtler than onion or garlic. Innumerable French recipes require this gourmet vegetable. The mild flavor blends well with parsley for meat sauces and can be used to enhance the flavor of eggs, salads and marinades. Shallots sauteed with mushrooms and seasoned with thyme and marjoram make an elegant side dish to accompany meats or fish.

But you need not wait for the bulbs to develop before reaping benefits from your shallot bed. Small amounts of the top growth can be cut at any time during the season, provided care is taken not to destroy the main, central shoot. These green leaves are chopped and used like chives, as seasoning and garnish. When young plants have reached the size of green onions, they may be harvested as fresh vegetables.

In whatever form the shallots are used, they are well worth the time, money, and garden space given them. The sets are expensive, but then it is a one-time investment that is returned, with capital, that very season. One pound of sets will return 5 to 7 pounds of shallots. About one-seventh of the harvest should be held over for replanting the following season. Most gardeners prefer to use or sell the large bulbs and keep the small ones for planting.

Shallot sets can be obtained from W. Atlee Burpee Co.; Casa Yerba; Comstock, Ferre and Co.; J.A. Demonchaux Co.; Gurney Seed and Nursery Co.; Le Jardin du Gourmet (who specialize in shallots, and carry several types); Johnny's Selected Seeds; J.W. Jung Seed Co.; Lakeland Nurseries; Earl May Seed and Nursery Co.; and Redwood City Seed Co. You can buy plants from Hemlock Hill Herb Farm and Well-Sweep Herb Farm.

• Shallots produce an additional bonus in that they can be useful to the cook as soon as the roots become established . . . the leaves of the shallot can be cut for seasoning and garnishes just like chives. As soon as the bulbs show signs of dividing, they can individually be removed from the group for kitchen use. Bulbs multiply in such a mass on healthy plants that often two or three can be lifted right off the top where they are held in the ground by five or six thin rootlets.

—*John Meeker*
Gilroy, California

Shungiku,
or Chop Suey Green

Chrysanthemum coronarium

IN the western world, this member of the Composite family is known as Garland Chrysanthemum and is grown for ornamental purposes in beds and borders. In China and Japan, however, it is called shungiku and is valued for its pungent, delicious leaves and shoots that are a major ingredient in such well-known dishes as chop suey and sukiyaki. The leaves of this edible chrysanthemum are fleshier than those of the ornamental type, and of a slightly different shape. If allowed to develop freely, it will reach a height of 2 feet, putting forth innumerable small, single yellow flowers.

Shungiku does best in cool weather and, like spinach, is usually grown as a spring and autumn vegetable. For a very early crop, plant seeds in the cold frame in February. Otherwise, plant outdoors as soon as the ground can be worked, sowing ¼ inch deep in ordinary, well-drained and fertile garden soil. When the young seedlings have developed their third set of leaves, thin them to stand about 4 inches apart.

Shungiku grows quickly and requires only clean cultivation and adequate watering to produce a bounteous harvest. Six to seven weeks after sowing, the leaves are about 5 inches long and can be gathered and eaten. You can either harvest the whole plant or cut only the tender side leaves at ground level. If it's not uprooted, shungiku will resprout.

• My wife set out the usual display of meat and sliced vegetables in preparation for making the well-known Japanese dish sukiyaki. I noticed that she had put flower stems on the plate. She had apparently snipped off the ends of our chrysanthemums, and I was not sure that I could eat them. She assured me however, that she had bought them in a Japanese market nearby and that they were a special chrysanthemum raised for eating. The explanation somehow satisfied me, and I enjoyed eating them. I found that they taste like chrysanthemums too, and though I cannot tell you what all chrysanthemums taste like, these—*Chrysanthemum coronarium*—I can assure you are good to eat. I now have them in my garden every season since I walked down the alley behind the grocery store where my wife first bought them and found that the grocery store owner grew his own out back for sale up front.

—*John Meeker*
Gilroy, California

Unless the plants are kept closely trimmed, they will flower in middle or late summer. Most gardeners find the leaves of the mature chrysanthemum rather stringy and their taste overwhelming. Nipping the flower buds before they have an opportunity to develop will help to maintain the tenderness and mild flavor of the leaves. But other gardeners enjoy the bit of color that the shungiku flowers add to the declining garden, and even claim that the leaves of the mature plant are as tasty as those of young ones. Certainly it is a good idea to allow a few plants to reach maturity so they can self-seed and provide new, early greens the following season. For a continuous supply of young shungiku leaves, sow a little seed every few weeks from February through early summer and from late summer through September. Summer sowings should be made in light shade to avoid bolting.

Rich in vitamin C, the greens are highly aromatic with a flavor resembling the smell of chrysanthemums. Cooking makes their pungent flavor somewhat milder, but many people enjoy them raw in salads and with tomatoes. The leaves can be stir-fried or steamed, like spinach, and served with butter. If it is not always enjoyed as a separate dish, most people will agree that a bit of shungiku enhances the flavor of soups, fish and meat dishes, and various Chinese and Japanese dishes.

Seeds of shungiku, or chop suey green, can be obtained from William Dam Seeds; DeGiorgi Co.; J.L. Hudson, Seedsman; Johnny's Selected Seeds; Kitazawa Seed Co.; Nichols Garden Nursery; Geo. W. Park Seed Co.; Redwood City Seed Co.; and Sunrise Enterprises.

Skirret

Sium sisarum

SKIRRET is one of the many hardy perennial vegetables which are typically grown as annuals. The name is derived from the Dutch *suikerwortel* meaning sugar root. In Scotland it is known as *crummock* and in Germany as *Zuckerwurzel.* A member of the carrot and parsley family, it is cultivated for the bunch of wrinkled, greyish roots that form from the crown. The roots have a sweet, tender white flesh which, when cooked like salsify or parsnips, is highly esteemed in oriental countries. The plant is native to China and Japan, but was also widely known to ancient Western civilizations. The Emperor Tiberius is said to have sought tributes of skirret roots from the warring Germans. During the fifteenth century it enjoyed a rather wide acceptance in Europe and, later, during the seventeenth and eighteenth centuries, it ranked as one of the major kitchen herbs and vegetables in English and American gardens. The great herbalist Gerard lists it as a very beneficial medicinal herb capable of curing everything from hiccups to frigidity. The seventeenth-century gardeners, Parkinson and Evelyn, provide their readers with recipes for boiled, stewed, and roasted skirret and praise it as one of the most acceptable and pleasant root vegetables, far surpassing the parsnip.

Yet, in spite of these worthy recommendations, skirret's popularity among Western gardeners proved to be only temporary; in the end, it lost out to the carrot, oyster plant, and, indeed, even to the parsnip. Most Western cooks and gardeners still look upon skirret as an inferior, old-fashioned vegetable, but the lasting traditions of the Eastern world continue to recommend it to creative gardeners and cooks.

In this country, skirret is usually raised from seed, but root cuttings are equally successful where available. Like horse-radish and sea kale, this plant produces thin, secondary side-roots which can be cut and planted. These cuttings must contain at least one bud or "eye" if propagation is to occur. They can be gathered and planted at any time.

Seed planting takes place in late March or April, or in the fall for spring crops. Sow seeds thinly in rows 18 inches apart. Skirret seeds are rather hard, making them slow to germinate. They will not germinate at all unless the soil is kept moist. Like most root crops, this one prefers a sandy loam, but it will also grow on heavier land. For the best roots, provide plenty of compost or manure and organic potash. Direct sunlight is most important to avoid insect damage and to encourage the full development of top growth.

When the seedlings have developed their fourth or fifth sets of leaves, thin them to stand about 9 to 12 inches apart in the rows. Apart from weeding and watering as necessary, skirret needs little attention during the growing season. If your soil is poor, it's a good idea to give your crop a midseason side-dressing of manure or compost.

Skirret responds well to interplanting with salad crops such as radishes, onions, and leaf lettuce. During the summer, the plants reach a height of about 2 feet, but since the foliage is delicate and rather sparse, it will not exclude light from interplanted crops.

If your skirret is planted early in spring, the roots will be large enough to harvest by October. Like salsify, skirret benefits from cool temperatures and tastes much sweeter after the first frost. The plants are very hardy and may even be left in the garden throughout the winter and harvested as needed. Or, dig them in late fall and store, like beets, in outdoor barrels, mounds, or a root cellar. The best temperature for storing skirret is just about 32°F. with a humidity of about 80 to 90 percent. Stored roots should not be allowed to freeze.

To prepare skirret for table, simply scrub the roots and cut them into suitable lengths for cooking. They can then be boiled with a bit of salt and served, like salsify or parsnips, with butter. The roots can be stewed, braised, baked, batter-fried, or creamed. They are also delicious when mashed with potatoes or served with a cheese sauce. Skirret can be grated or chopped and added to salads, or dressed in a vinegar marinade and served alone as salad. The flavor is described as very sweet and unusual, making this vegetable a fine ingredient in stews, curries, and soups.

Skirret plants are available from Hemlock Hill Herb Farm and Well-Sweep Herb Farm.

Glazed Skirret

Clean and cook a pound of skirret roots in boiling, salted water until tender, about 20 to 30 minutes. Drain, peel, and slice lengthwise. In a frying pan, melt 2 tablespoons of butter, then add a tablespoon of honey or molasses, ⅛ teaspoon nutmeg, and a sprinkling of powdered ginger or grated orange peel if you like. Add the skirret to the pan, stir to coat with the sauce, and cook until the roots are glazed and heated through. This recipe will serve 4 to 6, depending on your appetite.

Sorrel

Rumex, spp.

SORREL has a long history that reaches from the early British herbals to present day gourmet cookbooks. It has been used as a cure for scurvy, a blood cleanser, and a diuretic, as well as an elegant potherb. Its first cultivated form is said to have been introduced from France into Great Britain, where it rose to peak popularity in the sixteenth century. Mary, Queen of Scots, supposedly took some of the seeds to Scotland where it became a favorite among cooks. In Cornwall the herb was called sour sauce and was made into special sour-sauce pastries that are still baked today. But although sorrel is grown and appreciated in Europe, it has been generally ignored by American gardeners.

Known also as herb patience and patience dock, sorrel grows wild in the United States as well as in Europe. Sheep's sorrel is the most common wild form, and is prized by foragers as an especially tart and juicy salad stuff or cooked green.

The cultivated forms of sorrel are even tastier and, because they spread by clumps rather than runners, they can be more easily controlled in the garden bed. Two basic types are available to the American gardener: *Rumex acetosa* and *Rumex scutatus*. *R. acetosa* is the common garden sorrel known throughout the Near East and Europe. It is found growing wild as well as in gardens, is 12 to 36 inches tall, and has large, lance-like leaves. *R. scutatus,* or French sorrel, is the one that was so popular during Elizabethan times. Its leaves are smaller, and its height only 12 to 18 inches. Many gourmets prefer this type as its leaves are more acid and they make a tangier green vegetable. Both sorrels are rich in vitamin C. Both are hardy, perennial members of the buckwheat and rhubarb family. They have large, fleshy green leaves that form a rosette at the base of the plant. The flower stalk is long and slender, the flower pale yellow, and the seeds generally brown to red.

Whatever variety you choose, plant the seeds in a permanent bed where the soil is moderately acid. Sorrel prefers sun but will tolerate partial shade. The soil must be very rich in nitrogen and, as the roots are very deep, the plants should be watered liberally. Seeds can be sown outdoors in spring or fall, ½ inch deep or less in rows 18 inches apart. With enough water and nutrients, the plants will quickly fill in their allotted space.

Sorrel performs wonderfully with only a minimum of care. Once the plants have become established, pick the outer leaves of the rosette, as you would parsley, to encourage the branches to spread well. Supply abundant nitrogen with side-dressings of rotted manure or compost each spring and if your soil is in really bad shape, monthly feedings of fish emulsion throughout the season. Pick off the older, yellowed leaves and cut the seed stalks as they appear in midseason. This will direct the plants' energy into the production of edible leaves and prevent unwanted seeds from scattering over the garden. If left unmanaged, sorrel can become an impossible weed.

To harvest sorrel, cut or pinch off the outer leaves, always being careful to leave a rosette of young growth to insure future harvests. Regular harvesting, even if you don't use the produce, will keep the crop at its best and most tender.

Sorrel wilts very quickly after harvest (which is one reason you so seldom see it in stores), so cut the leaves right when you're ready to use them. Pick over the leaves, removing the tough parts of the stems; then rinse and drain. Sorrel can be prepared in any number of ways. Its culinary uses include salad greens, and most famous, cream of sorrel soup. The tangy greens are a favorite served with fish, particularly trout or salmon. It also figures as an ingredient in Old English

• My introduction to the sorrels was not under cultivation, but in the wild garden along California's north coast. Sheep's sorrel grows abundantly in clearings and meadows in the moister areas, and I used to pick the lance-shaped leaves to pep up otherwise bland domestic salads, or to nibble the thirst-quenching leaves on hikes. As my gardens developed, I found that sheep's sorrel (*Rumex acetosella*) responds admirably to organic methods.

In the wild I rarely found leaves growing over 3 inches but given garden conditions they volunteered at 8 and 9 pease porridge. For an added zip in cooked chard or spinach, a few leaves of sorrel can be mixed in.

Medicinally, the fresh leaves can be used to soothe canker sores or nettle stings. A tea made from the leaves is diuretic. A decoction of the fresh root has astringent properties and is useful for stomach hemorrhage and allied problems.

Rarely bothered by insects or diseases, a crop of sorrel will last for several years, rejuvenating itself continuously except during cold winter months. Plants set out new roots and crowns so that each plant gradually becomes a clump. If after several years they become too crowded, turn the plants under and sow new seed, or divide these clumps and set the plants in newly tilled, fertilized soil where they will resume bearing in a month or two. For the average family, eight or ten plants will prove sufficient. If the crop begins to get out of hand, a small dose of lime will discourage and control further growth.

Sources of seed for sorrel include Casa Yerba; Comstock, Ferre and Co.; DeGiorgi Co.; J.A. Demonchaux Co.; Charles C. Hart Seed Co.; J.L. Hudson, Seedsman; Le Jardin du Gourmet (where you'll find it listed under its French name, *oseille*); Johnny's Selected Seeds; Meadowbrook Herb Garden; Mellinger's, Inc.; Nichols Garden Nursery; Redwood City Seed Co.; R.H. Shumway Seedsman; and Thompson and Morgan, Inc. Hemlock Hill Herb Farm has plants, and Well-Sweep Herb Farm offers both seed and plants.

inches—very tender and tasty. I found that I had a perfect no-work perennial vegetable on hand—except for one fact I'm certain many other gardeners across the country have learned about this plant: Sheep's sorrel propagates by runners and it's not long before intertwining sorrel root systems are interfering with valuable cultivated crops. I found it especially hard to deal with among the strawberries because even the tiniest piece I missed in weeding seemed to have the power to reinfest the whole patch.

—*James Jankowiak*
Eureka, California

Sorrel Sauce

Here's how to make a simple sorrel sauce to serve with fish. Clean a pound of sorrel and put it in a saucepan along with ½ cup of water, 2 tablespoons of butter, and salt and freshly ground pepper to taste. Cook until the sorrel is wilted, about 5 to 10 minutes. Puree the greens and liquid in a blender, return to the pan, and add a few tablespoons of cream. Heat through, but don't allow the sauce to boil. This recipe should make enough sauce to accompany 4 or 5 servings of fish.

New Zealand Spinach

Tetragonia expansa

NATIVES of the Pacific islands and Australia had been eating *Tetragonia expansa* for centuries before Sir Joseph Banks "discovered" the plant along Queen Charlotte's Sound in New Zealand. The explorer brought back specimens to England where, as "New Zealand spinach," it enjoyed a modest popularity among *avant garde* eighteenth century gardeners. Today it is one of the best and most widely cultivated summer greens. Heat and drought resistance, high yields, and an ability to withstand cultural abuse make it an excellent spinach substitute for summer months. Unlike spinach, it does not bolt or become bitter in warm weather and, when only the growing tips are harvested, it supplies greens continuously from May through October.

When cooked, New Zealand spinach is practically indistinguishable from true spinach but, horticulturally, it is completely unrelated. A member of the carpetweed family (*Aizoaceae*), it is a very vigorous, low-growing, and spreading annual with thick triangular leaves. The small yellowish green flowers are followed by hard pods that shed seeds in the fall and put forth sometimes unwelcome volunteers the following spring.

The plant is very sensitive to frost, yet is able to withstand some cooler temperatures. In maritime areas it will survive the winter, but most gardeners feel that optimum quality is

• I became a grower the first season a group of us developed a community garden in the part of northern coastal California where I then lived. A neighbor suggested we grow it as a cash crop to defray equipment expenses of our new garden. He guaranteed it was a vegetable that grew from spring to frost, supplied market-ready greens continuously from the age of two months, provided its own living mulch, and required no care other than watering and harvesting. The amazing thing is that New Zealand spinach performed as promised. It also tasted good, and thereby earned a place in my garden as one of the regulars. What my neighbor didn't tell me was that it's a reliable self-sower. When the ground warms up in spring, seedlings spring up in the old bed, ready to be transplanted to new rows.

—*James Jankowiak*
Eureka, California

achieved in strictly summer crops. Although it is a warm weather crop, New Zealand spinach is unable to grow in the extremely hot climate of the Deep South unless protected from the direct rays of the sun. It will thrive in greenhouses if given plenty of water, but the most worthwhile way to raise the crop is outdoors, during the summer months in temperate regions. The seeds should be started indoors or under glass around February. For quick germination, you can soak them overnight in cool water (three or four hours in hot water), and sow in flats, two to three seeds to the inch, ½ inch deep. Transfer very small seedlings to peat pots and plant outdoors after all danger of frost has passed. Set plants in the garden at 1- to 2-foot intervals in rows spaced 3 to 4 feet apart. A very short row will supply the needs of an average family. Since it takes nearly two months for the plant to fill that area, leaf lettuce, radishes, carrots, or some other short-season crop can be planted between rows. These will be out of the way by the time the spinach requires the space.

If you prefer direct-seeding, sow your pre-soaked seeds in open rows late in the spring. Thin or transplant the seedlings, as needed, with a ball of soil attached to the roots.

New Zealand spinach likes fertile, deeply cultivated and well-drained soils in a sunny location. Its lush growth demands plenty of compost or well-rotted manure and, if the soil is a bit acid, an application of lime is helpful. The pH should be about 6.5 or 7.

Although it leaves store amounts of moisture sufficient to resist drought, the plant performs best if watered liberally whenever the top inch of soil is dry. Plenty of nitrogen and water will make for the fastest-growing, and therefore the tenderest, leaves and stems.

When the main runners begin to spread, pinch them back to encourage more growth. A dry mulch of sawdust can be applied at this time, but is not absolutely necessary as the plants form their own dense carpet.

There are basically two ways to harvest New Zealand spinach. Commercial growers usually harvest all of the top growth two or three times each summer. The plant is held in one hand and the stems cut 2 or 3 inches from the ground.

Home growers, on the other hand, generally prefer to harvest the spinach by the "cut-and-come-again" method. By this system, 3- to 4-inch tips are snipped from the branches as needed. The supply is neverending and, as only the edible portion is harvested, there is very little waste in preparing this vegetable for table. It is a good idea to pick these growing tips at least once a week whether you need them or not. This helps to encourage a continuous supply of tender, young leaves. The older leaves are tough and bitter-tasting and should be avoided, along with the flowers and pods.

Because the growing tips are generally upright, the edible leaves do not come in contact with the soil and there is no sand to be washed off. The plant's leaves are smooth and will not harbour insects. In fact, New Zealand spinach seems to be ignored by most insect pests and diseases. Slugs and snails that might bother other greens are uninterested in this one.

The leaves and young shoots can be used in place of spinach in any recipe and are much easier to handle. They can be steamed, boiled, stir-fried or made into a quiche. Try them creamed, served with mushrooms, or prepared au gratin. The flavor is very similar to that of spinach, so much so that most people can't tell the difference. The very young shoots can be eaten as salad, but most gardener-cooks seem to find the larger leaves rather disagreeable unless cooked.

New Zealand spinach seeds are available from many suppliers, including W. Atlee Burpee Co.; D.V. Burrell Seed Growers Co.; Comstock, Ferre and Co.; William Dam Seeds; DeGiorgi Co.; Farmer Seed and Nursery Co.; Gurney Seed and Nursery Co.; Joseph Harris Co., Inc.; Charles C. Hart Seed Co.; H.G. Hastings Co.; Jackson and Perkins Co.; Le Jardin du Gourmet; J.W. Jung Seed Co.; Earl May Seed and Nursery Co.; Mellinger's, Inc.; Nichols Garden Nursery; L.L. Olds Seed Co.; Redwood City Seed Co.; Seedway; R.H. Shumway Seedsman; Spring Hill Nurseries; and Stokes Seeds, Inc.

Spaghetti Squash

Cucurbita pepo

MANY'S the gardener who, out of curiosity, plants a hill of spaghetti squash, only to find it is delicious and wish he had planted more. An egg-shaped squash, it has stringy flesh that really does look like spaghetti, and is delicious when served hot with spaghetti sauce, cheese, or butter. A fully ripe spaghetti squash is yellowish, and measures 8 to 12 inches in length with a diameter of about 9 inches. This vegetable is also known as vegetable spaghetti and Manchurian squash, and was introduced to this country in the early 1930s by a Japanese seedsman. Today, it is offered by a host of American suppliers.

The culture of spaghetti squash is identical to that of most other squashes. Like cucumbers, it performs well in the greenhouse when the vines are trellised or trained to climb up string supports. Outside it requires full sun, plenty of room to spread, and a soil rich in humus. Very sandy soils should be enriched with compost, well-rotted manure, or a green manure crop for the best results. Garden beds adequate for other cucurbits will easily support spaghetti squash.

Plant seeds in a sunny, well-drained location well after the last expected frost date. Although they can be planted 8 inches apart in rows, most gardeners prefer to plant their vegetable spaghetti in hills. Loosen the soil to a depth of 1 foot, place a shovelful of compost, some granite dust, and, if necessary, a

• In early fall, we placed several large squash in a gunnysack with a bunch of butternut and acorn squash, and stored the whole thing under the house after cool weather set in. Much to our surprise, the spaghetti squash was still firm, bright and delicious even after the acorns and butternuts had begun to soften and shrivel.

—William F. Rau and
Richard T. Mitch
Dunlap, California

377

touch of lime, beneath each hill, then mound the dirt about 4 inches high. Hills should be spaced 4 to 6 feet apart with four or five seeds sown in the top of each one. Plant the seeds ½ inch deep and 2 inches apart, then water thoroughly.

Since a half ounce of seeds plants about twelve hills, you'll probably have some seed left over. Store these in a tightly covered jar for use next year. If properly stored in a cool place (like a basement), they will remain viable for several years.

When the young plants are 3 inches high, thin them to two or three per hill. Cultivate the hills lightly to remove weeds and mulch heavily with clean, dry hay, straw, or sawdust. Except for a dose of compost or manure tea just before the blossoms open and some extra water during dry periods, your crop of spaghetti squash will require little care and feeding after this.

Pests may be a problem, however. The same insects that attack cucumbers and other squashes can easily do in a few hills of spaghetti squash. To deter many of the leaf-eating beetles, dust the plants liberally with wood ashes. Not only does this practice repel many insects, but it also adds potash to the soil. Hand-pick squash bugs as they appear and, where insects are a real menace, spray or dust with rotenone. Borers are best controlled by clean cultivation, crop rotation, and the elimination of diseased plants.

Vegetable spaghetti plants are very lush-growing, producing a large number of fruits on each vine. The squashes are best if picked 90 to 110 days after seeding, when the skins have turned from pale green to a light yellow or cream color. In August, the squashes generally begin to change color and by September they are ready to harvest. At this time the stem should be rather dry and the fruit should pull easily away from the vine. If frost threatens before the fruits have ripened, pick them while still green and place in a cool, not cold, dry place where they will yellow. Or, fry the green ones in oil and serve them like eggplant.

To prepare mature spaghetti squash, boil them for 20 minutes to a half hour until the firm outer shell softens. Cut the fruit lengthwise, remove the center pulp and seeds, and rake out the strands of flesh. Or, simply prick the raw squash with a

I would plant this vegetable even if I didn't like it mainly for the unusual quality of it. It is unbelievable to find how much of the strings of 'spaghetti' there is in just one fruit. It will fill a big platter. We like it with butter, salt and pepper, as well as with regular spaghetti sauce. It has a unique "crunch" when you bite into it or you would not be able to distinguish it from the one made from wheat. I always enjoy seeing the expression on peoples' faces who have never seen spaghetti squash before, and I admit that it is enjoyable for me even after this long to watch it come to life. I have never had a failure with it— and that is a good recommendation!

—*R.L. Brewer*
Hampton, South Carolina

fork and bake at 350°F. for an hour and a half. Then cut it open and remove the flesh. A large squash will contain tremendous amounts of spaghetti-like strands. Serve it, while still warm, with salt, pepper, and butter; with garlic, a light cheese sauce, or mushroom sauce. The flavor is very delicate and slightly sweet, somewhat akin to that of acorn squash. Eaten with an Italian tomato and meat sauce, it is a delightful, low-calorie substitute for the usual pasta. Or try your usual lasagna recipe, using spaghetti squash instead of noodles. Chilled, vegetable spaghetti makes an intriguing addition to a tossed green salad. It's even been used, along with coconut flavoring, in cream pies.

Success with the storage of this rather thin-skinned squash is uncertain. Some gardeners have found that they keep as well as the heavier winter squashes such as acorn and butternut, when stored in a cool, dry spot. Others prefer to cook all their spaghetti squash soon after harvest and store it in the freezer.

Seeds for spaghetti squash can be purchased from many suppliers, including Burgess Seed and Plant Co.; W. Atlee Burpee Co.; Comstock, Ferre and Co.; DeGiorgi Co.; Henry Field Seed and Nursery Co.; Grace's Gardens; Gurney Seed and Nursery Co.; J.L. Hudson, Seedsman; J.W. Jung Seed Co.; Lakeland Nurseries; Earl May Seed and Nursery Co.; Nichols Garden Nursery; L.L. Olds Seed Co.; Geo. W. Park Seed Co.; Redwood City Seed Co.; Seedway; R.H. Shumway Seedsman; Stokes Seeds, Inc.; and Thompson and Morgan, Inc.

Vegetable Spaghetti In Our Greenhouse

It may seem strange, but you can actually grow "spaghetti" in the greenhouse. We're referring to vegetable spaghetti, a type of squash that grows 8 to 10 inches long, and turns yellow when mature.

You can plant the seeds directly in the bench, or use pots if space is a problem. Sow seeds about 8 inches apart and as soon as they are up, better make arrangements for the vines to crawl up strings. Keep the plants watered, and feed fish emulsion about once every 3 or 4 weeks. As soon as the fruits are yellow, you'll know they are mature—time to harvest them. Boil whole fruits for about 20 to 30 minutes, after which the flesh can be raked out in long, spaghetti-like strands. Season with salt, pepper and butter, or if you wish, use regular spaghetti sauce.

— Doc and Katy Abraham
Naples, New York

Tomatillo

Physalis ixocarpa

IF your tastebuds are partial to Mexican dishes, you should by all means consider growing the tomatillo. This first cousin of the ground cherry is the basic ingredient in *salsa verde*, the mildly hot green sauce often served with tacos, *chilis rellenos,* and other Mexican dishes. The sticky green berries generally grow to about the size of walnuts. Their flavor is less sweet and not as pronounced as the taste of ground cherries, and reminds some people of green tomatoes. But tomatillos and green tomatoes are never interchangeable in recipes—the flesh of the tomatillo is of a different texture, seedy but solid, without the juicy cavities found in tomatoes.

A native of Mexico, the tomatillo can be grown in outdoor gardens in warm parts of the United States, and in the greenhouse further north. On the West Coast, the fruit can be found in supermarkets, but in most parts of the country, the only way to get it is to grow it yourself. Tomatillos are grown like ground cherries or tomatoes, and they are generally not too difficult to grow. In fact, one Oregon gardener raised a crop in a dry, rocky corner of her garden where other plants refused to grow.

Plants are generally started indoors, on a sunny windowsill. They'll be ready for transplanting in a month, either into individual pots or into the garden. Be sure all danger of frost is

• Here's a simple method of extracting tomatillo seeds from the fruit for planting, suggested by Bauman's Pickle Room: Peel the husk, wash fruit, and skin with a knife to a depth of ⅛ to ¼ inch. There are no seeds in this outside layer. Place the "heart" in a blender with about a cup of water. Blend until the pulp is finely chopped. The viable seeds will sink to the bottom, while the chaff will float. Several rinsings will leave only the seed.

Drain excess water and the seeds are ready to plant. They do not need to be dried in order to germinate. They may be dried in an oven for storage, but do not let the temperature get above 110 degrees for even a second.

381

past and the soil is warm before setting out the plants. Space the seedlings a foot apart in the garden, and take care to keep them well watered and protected from the sun for the first few days. After that, a layer of mulch and an occasional watering are about all the care your tomatillos will need until the fruits mature in fall.

For use in cooking (and for the best flavor), the fruits should be picked while they are still deep green, and when the husk has changed from green to tan. If left on the vine to ripen, tomatillos will turn either yellow or purple, and develop a sweet, bland taste. They may be used in pies and preserves when ripe, but their blandness makes them less well suited to the purpose than ground cherries.

Each plant should yield about a pound of fruit. Tomatillos keep well in a cool, well-ventilated place, stored one layer deep, still in their husks. This type of storage may even keep them through the winter. Jammed into airtight plastic bags, though, the berries spoil rapidly.

Excess fruits may also be canned. Remove the husks and wash well, then cook whole tomatillos in boiling water, covered, until tender (about 10 minutes). Drain and pack into hot jars, and fill with boiling water to within ½ inch of the rim. To each pint, add ½ teaspoon of salt and 1 teaspoon of vinegar or lemon juice. Seal and process in a boiling water bath for 30 minutes.

These canned tomatillos may be used in making *salsa verde*, or you can simply cook down the fresh fruit. An Iowa gardener recommends slicing tomatillos raw in a salad, along with tomatoes, cucumbers, and onion. Dress with your favorite vinaigrette sauce.

An Oregon gardener reports the following meatless meal is a big hit at her house, especially with her kids. Spoon hot refried beans into crisp, cooked taco shells. Top with fresh garden vegetables (lettuce, tomatoes, green peppers, onions, celery, cucumbers—whatever is on hand), shredded cheddar or Monterey Jack cheese, and homemade green taco sauce. (See recipe).

Tomatillo seed is available from Bauman's Pickle Room and Horticultural Enterprises.

Tomatillo Dip

Blend 1 cup chopped fresh tomatillos, ½ cup chopped onion, 2 tablespoons dried coriander leaves, ½ to 1 tablespoon seeded, minced jalapeno pepper, and ½ teaspoon garlic salt in a blender until smooth. Mix with ½ cup sour cream, chill and serve with tortilla or corn chips. This recipe makes about 2 cups of dip.

— Bauman's Pickle Room
Spring Valley, California

Green Taco Sauce
(*salsa verde*)

1 pint canned or cooked
  tomatillos, drained
¼ cup fresh coriander leaves,
  chopped
2 cloves garlic
1 small onion, coarsely
  chopped
½ teaspoon salt
1 tablespoon lime or lemon
  juice
  chopped green chile peppers, fresh, canned or
  pickled. Use your own
  judgement here. How hot
  do you like it? How hot are
  your chiles? For canned
  California green chiles, try
  4 tablespoons and work up.

Combine ingredients in
blender jar and blend until
smooth. This sauce may be
frozen.

—Polly Timberman
Canyonville, Oregon

Appendix 1

Seed Sources for The Vegetables in This Book

Bauman's Pickle Room
P.O. Box 628
Spring Valley, CA 92077
   Their specialty is pickles, but they supply hard-to-find tomatillo seeds.

Burgess Seed and Plant Co.
P.O. Box 3001
Galesburg, MI 49054
   A good selection of basic vegetables, especially those suited to cooler climates, and several unusual items as well.

W. Atlee Burpee Co.
Clinton, IA 52732
Warminster, PA 18974
   Large, well-known company with a wide selection of most vegetables and flowers.

D. V. Burrell Seed Growers Co.
Rocky Ford, CO 81067
   A good selection of basic vegetables, especially melons and chile peppers.

Casa Yerba
Star Route 2, Box 21
Days Creek, OR 97429
   A wide selection of herb seeds and plants.

Chase Compost Seeds, Ltd.
Benhall, Saxmundham
Suffolk, England
   Organic vegetable and flower seeds, English and European varieties. They carry four kinds of runner beans.

Comstock, Ferre and Co.
263 Main St.
Wethersfield, CT 06109
   Old house with excellent selection of hard-to-find vegetables and a nice assortment of herbs.

William Dam Seeds
Highway 8
West Flamboro, Ontario
Canada L0R 2K0
   Canadian company with a wide selection of vegetables and herbs. They carry many European varieties.

DeGiorgi Co.
P.O. Box 413
Council Bluffs, IA 51501
   One of the best selections of vegetables, including old-fashioned and unusual ones.

J. A. Demonchaux Co.
225 Jackson
Topeka, KS 66603
  Specializes in gourmet vegetables, and stocks several varieties of vegetables such as sorrel and corn salad, which are popular in Europe.

Exotica Seed Co.
820 South Lorraine Blvd.
Los Angeles, CA 90005
  This small firm specializes in tropical and subtropical trees, fruits, and ornamentals. They carry jicama and prickly pear cactus, which are difficult to find.

Farmer Seed and Nursery Co.
Faribault, MN 55021
  A small house carrying trees, flowers, and a nice selection of vegetables primarily suited for cool climates.

Henry Field Seed and Nursery Co.
Shenandoah, IA 51602
  Mostly flowers, trees, and fruit but handles some unusual vegetables.

Glecklers Seedmen
Metamora, OH 43530
  Carries some uncommon vegetables, most notably two species of basella.

Grace's Gardens
22 Autumn Lane
Hackettstown, NJ 07840
  Small company specializing in unusual vegetables, oriental vegetables, dwarf and jumbo varieties.

Gurney Seed and Nursery Co.
Yankton, SD 57078
  A varied assortment of unusual vegetables, cold-weather vegetables, and fruit trees.

Joseph Harris Co., Inc.
Moreton Farm
Rochester, NY 14624
  A good selection of basic vegetables, and a substantial number of the vegetables in this book.

Charles C. Hart Seed Co.
304 Main St.
Wethersfield, CT 06109
  A diversified selection of old-fashioned and non-hybrid vegetables.

H. G. Hastings Co.
P.O. Box 4274
Atlanta, GA 30302
  Specializes in southern fruits, vegetables, and flowers.

Hemlock Hill Herb Farm
Hemlock Hill Road
Litchfield, CT 06759
  A small firm that supplies several hard-to-find plants, including comfrey, burnet, skirret, and Egyptian onion. Catalog 50 cents.

Horticultural Enterprises
P.O. Box 34082
Dallas, TX 75234
  Their specialty is chile peppers, and they have the best selection we've seen. They also supply jicama, tomatillo, and coriander.

J. L. Hudson, Seedsman
P.O. Box 1058
Redwood City, CA 94064

Call themselves "A World Seed Service," and indeed they are. Among the best selections of unusual vegetables, carrying such hard-to-get items as sea kale, jicama, scorzonera, and martynia. They also carry a remarkable collection of flowers, herbs, trees, and ornamentals from around the world. Catalog, unillustrated, 50 cents.

Jackson and Perkins Co.
Medford, OR 97501

Lots of flowers, and a nice selection of vegetables, including a fair number of unusual ones.

Le Jardin du Gourmet
West Danville, VT 05873

Their specialty is shallots, and they carry several varieties. True to the name, the emphasis here is on gourmet vegetables. An interesting and varied selection of unusual vegetables, including most kinds of onions, a good variety of greens, European favorites, and traditional American crops as well. They also import Vilmorin seeds from France.

Johnny's Selected Seeds
Albion, ME 04910

Small seed company with integrity. Carries native American crops, select oriental vegetables, grains, and fast-maturing soybeans for northern areas. Catalog 50 cents.

J. W. Jung Seed Co.
Randolph, WI 53956

A small but varied selection of the vegetables in this book. They carry horseradish crowns, which aren't always easy to find.

Kitazawa Seed Co.
356 W. Taylor St.
San Jose, CA 95110

A small West Coast firm carrying only oriental vegetables, particularly Japanese varieties.

Lakeland Nurseries
Hanover, PA 17331

Primarily fruits, flowers, and trees, but several of the vegetables in this book as well.

Earl May Seed and Nursery Co.
Shenandoah, IA 51603

Again, the emphasis here is on flowers and fruit, but they also have a good selection of vegetables, and an assortment of those in this book.

Meadowbrook Herb Garden
Rt. 138
Wyoming, RI 02898

Another source for hard-to-find herbs and traditional vegetables, such as nettle, Egyptian and Welsh onions, and orach. Biodynamically grown spices, herbs, teas, and herb seeds.

Mellinger's, Inc.
North Lima, OH 44452

This firm has a large selection of trees, but offers quite a few unusual vegetables as well.

Nichols Garden Nursery
1190 North Pacific Highway
Albany, OR 97321
  An interesting and unusual selection of
  vegetables and herbs, including many of
  the vegetables in this book.

L. L. Olds Seed Co.
P.O. Box 7790
Madison, WI 53707
  This firm carries basic vegetables, and a
  substantial number of unusual items as
  well.

Geo. W. Park Seed Co.
Greenwood, SC 29647
  The emphasis here is on flowers, and
  their catalog is beautiful; they also carry a
  diverse selection of vegetables.

Redwood City Seed Co.
P.O. Box 361
Redwood City, CA 94061
  A well-rounded selection of non-hybrid,
  untreated vegetable and herb seeds,
  including many of the vegetables in this
  book. They're also expert on locating
  various tree seeds.

Seedway
Hall, NY 14463
  A nicely balanced assortment of vegetables, including several found in this
  book.

R. H. Shumway Seedsman
Rockford, IL 61101
  An old house with a charming turn-of-
  the-century style catalog. Many varieties
  of standard vegetables, plus quite a few
  unusual items. Also a good selection of
  grains, fodders, and cover crops.

Spring Hill Nurseries
110 W. Elm St.
Tipp City, OH 45371
  A small selection of unusual vegetables.
  They carry five varieties of sweet potato
  plants.

Stokes Seeds, Inc.
Box 548
Buffalo, NY 14240
  Carries excellent varieties of many vege-
  tables, especially carrots.

Sunrise Enterprises
P.O. Box 10058
Elmwood, CT 06110
  A good selection of oriental vegetables,
  particularly Chinese varieties.

Thompson and Morgan, Inc.
P.O. Box 100
Farmingdale, NJ 07727
  This British firm carries a wide selection
  of vegetables, including many that are
  hard to find, like asparagus pea and
  jicama.

Tsang and Ma International
P.O. Box 294
Belmont, CA 94002
  A small house specializing in Chinese
  vegetables; they offer a good selection.

Otis S. Twilley Seed Co.
Salisbury, MD 21801

  A basic assortment of standard vegetables, oriented toward warm-climate
  growing.

Vermont Bean Seed Co.
Ways Lane
Manchester Center, VT 05255
  As their name implies, a fine selection of beans is available from this small firm.

Well-Sweep Herb Farm
317 Mt. Bethel Rd.
Port Murray, NJ 07865
  An excellent selection of herb plants and seeds, scented geraniums, and dried flowers. They have such hard-to-find plants as Good King Henry, orach, and skirret.

Appendix 2

Vegetable	Annual Biennial Perennial	Planting Time
Amaranth	A	Early spring, after most danger of heavy freezing has passed. In subtropical regions, plant in early fall. Transplants: Where growing season is short, start indoors or under glass 8 weeks before last expected frost.
Bamboo	A, P	Late spring, well after danger of frost has passed and the soil has warmed.
Basella	A	Late spring, well after danger of frost has passed and the soil has warmed.
Asparagus Bean	A	Late spring, well after danger of frost has passed and the soil has warmed.
Fava Bean	A	Early spring, as soon as the ground can be worked and danger of extreme cold has passed. In mild, coastal areas, plant in late fall or early winter.
Horticultural Bean	A	Late spring, well after danger of frost has passed and the soil has warmed.
Purple Bush Bean	A	Late spring, after danger of frost has passed and the soil has warmed.
Scarlet Runner Bean	A, P	Late spring, after danger of frost has passed and the soil has warmed to 55 to 60°F. Transplants: Start indoors or under glass several weeks before last expected frost.

Vegetable	Annual Biennial Perennial	Planting Time
Soybean	A	Late spring, after danger of frost has passed.
Borage	A*	Early spring, as soon as the ground can be worked. For a continuous crop, plant at two-week intervals through midsummer.
Asian Brassicas		
Chinese Cabbage	A	Late spring to early summer, well after danger of frost has passed.
Michihli	A	Early to midsummer; not recommended before July in most areas.
Bok Choy	A	Early spring; late summer.
Broccoli Raab	B	Early spring, as soon as the ground can be worked; fall.
Domestic Burdock	A, B	Late spring, well after danger of frost has passed.
Salad Burnet	P, A	Early spring, about the time of the last frost.
Nopal and Prickly Pear Cacti	P	Anytime except winter. Transplants: Start indoors or under glass.
Cardoon	P, A	In mild areas, direct-seed in late spring, after danger of frost has passed; elsewhere set out plants after soil has warmed. Transplants: Start indoors or under glass about 8 to 10 weeks before soil warms.
Purple Cauliflower/ Purple Broccoli	A	Late spring, well after danger of frost has passed. Transplants: Start indoors or under glass, 6 to 8 weeks before last expected frost.
Celeriac	A, B	In mild areas, direct-seed in late winter. Elsewhere, start indoors and set out in early summer. Transplants: Start indoors or under glass about 8 weeks before last expected frost.

*indicates vegetables which reseed themselves.

Vegetable	Annual Biennial Perennial	Planting Time
Celtuce	A, B	Early spring, as soon as the ground can be worked; late summer, for a fall crop.
Chayote	A, P	Late spring, after danger of frost has passed.
Chicory	A, P	Non-forcing types: early to midsummer. Forcing types: late spring, after danger of frost has passed, to early summer. In cool regions, forcing chicory should be started indoors; in warm areas, non-forcing types should be sown in early summer.
Collards	A, B	Midspring through early summer, after most danger of extreme frost is past. Late fall in mild regions.
Comfrey	P	Spring, as soon as the ground can be worked and most danger of frost has passed.
Coriander	A	Early spring.
Corn Salad	A	Early spring, before danger of frost has passed; late summer for fall use. In mild regions, sow in fall for winter/spring use.
Cowpea	A	Late spring, after danger of frost has passed.
Winter and Garden Cresses	P, A	Very early spring, as soon as the ground can be worked. For a continuous supply, sow every three weeks until midsummer. In mild areas, sow seeds July through early fall, for winter use.
Watercress	P	Very early spring, as soon as the ground can be worked.
Chinese Cucumber	A	Late spring, well after danger of frost has passed and the soil has warmed.
Dandelion	A*, P	Very early spring, as soon as the ground can be worked; midsummer to early fall, for winter and early spring greens.
Daylily	P	Anytime but winter; clumping varieties divided and replanted in the fall.
Japanese Eggplant	A	Transplants: Start indoors or under glass 10 to 12 weeks before last frost.

Vegetable	Annual Biennial Perennial	Planting Time
Florence Fennel	A, P	In temperate regions, plant in late spring, after most danger of frost is past; in milder regions, plant in late fall, autumn, or winter.
Garlic	A, P	Very early spring, as soon as the ground can be worked. In mild areas, plant in fall.
Elephant Garlic	A, P	Very early spring, as soon as the ground can be worked; late summer or fall.
Good King Henry	A*	Very early spring, as soon as the ground can be worked.
Ground Cherry	A	Early summer, after the soil has warmed and all danger of frost has passed. Transplants: Start indoors about 8 weeks before last expected frost.
Horseradish	P	Early spring, as soon as the ground can be worked; late fall.
Jerusalem Artichoke	P, A	Early spring.
Jicama	A	Late spring, as soon as soil is thoroughly warm, and all danger of frost has passed.
Kale	A, B	Very early spring, as soon as the ground can be worked; also midsummer. In mild regions, plant in fall.
Flowering Kale	A	Early spring, as soon as the ground can be worked. Transplants: Start indoors or under glass 6 to 8 weeks before soil has warmed.
Kohlrabi	A, B	Early spring to mid or late summer, successive sowings. Transplants: Start indoors, 4 to 6 weeks before last expected frost.
Leek	P, A	Early spring, as soon as the ground can be worked, until early summer. Transplants: Start indoors or under glass, about four weeks before soil can be worked.

Vegetable	Annual Biennial Perennial	Planting Time
Luffa	A	Late spring, after all danger of frost has passed. In temperate regions, start indoors. Transplants: Start several weeks before last expected frost.
Martynia	A*	In very mild areas, direct-seed in spring, after all danger of frost has passed. Elsewhere set transplants outdoors in early summer. Transplants: Start indoors or under glass about 8 or 10 weeks before soil has warmed.
Mustard	A	Early spring, as soon as the ground can be worked; successive plantings every few weeks until early fall.
Mustard Spinach	A	Early spring, before last frost until fall.
Nasturtium	A	Early spring, after danger of frost has passed.
Nettle	P*	Early spring as soon as the ground can be worked; late summer or fall for spring crop.
Okra	A	Early summer, well after danger of frost has passed and soil has warmed.
Egyptian Onion	P	Early spring, as soon as the ground can be worked.
Welsh Onion	P	Late summer.
Orach	A	Early spring, after the danger of frost has passed; early autumn.
Hamburg Parsley	A, P	Early in spring, as soon as the ground can be worked.
Asparagus Pea	A	Late spring, after ground has warmed.
Sugar Pea	A	Early spring, as soon as the ground can be worked.
Peanut	A	Late spring to early summer.
Popcorn	A	Late spring to early summer, as soon as the soil has warmed to above 50°F.
Sweet Potato	A	Sprouts planted in late spring or early summer, about 10 days after last expected frost.
Lady Godiva Pumpkin	A	Late spring, as soon as danger of frost has passed and soil has warmed.

Vegetable	Annual Biennial Perennial	Planting Time
Purslane	P	Early summer through late summer.
Winter Radishes	A	Midsummer, in temperate regions; late summer in milder areas.
Rhubarb	P	Seeds sown after soil has warmed and all danger of frost has passed; crowns planted in early spring as soon as ground can be worked or in late fall after the first frost.
Rocket	A, P	Early in spring or summer, after most danger of frost has passed. In mild regions, plant in late fall for a spring crop.
Salsify	A, B	Early in spring, as soon as the ground can be worked; early summer in warm regions. Or plant in fall, for spring use.
Scorzonera	A, B	Early spring, as soon as the ground can be worked.
Sea Kale	A, B	Early to late spring, as soon as the ground can be worked; autumn cuttings can be planted if protection is provided.
Shallot	A, P	Early spring, as soon as the ground can be worked, through late spring.
Shungiku	A*	Early spring, as soon as the ground can be worked; late summer. At two- or three-week intervals for a continuous supply of greens.
Skirret	A, P	Early spring, as soon as the ground can be worked; fall for a spring crop.
Sorrel	P	Early spring, as soon as the ground can be worked, until early fall.
New Zealand Spinach	A	Late spring, after danger of frost has passed, through midsummer.
Spaghetti Squash	A	Early summer, after danger of frost has passed and the soil has warmed, until midsummer. Transplants: Start indoors or under glass about 6 weeks before soil has warmed.
Tomatillo	A	Transplants: Start seeds indoors or under glass about 4 weeks before last expected frost. Set outside when soil has warmed.

Appendix 3

Vegetable	Soil Type	Nutrition
Amaranth	Well-tilled, rich, but will tolerate poor soil.	Responds well to added phosphorus and nitrogen
Bamboo	Sandy, well-drained; very acid	
Basella	Sandy, well-drained; pH 6.0 to 6.7	High nitrogen requirements
Asparagus Bean	Ordinary	Prefers soil slightly deficient in nitrogen.
Fava Bean	Ordinary; pH 5.5 to 6.7	Prefers soil slightly deficient in nitrogen
Horticultural Bean	Ordinary; not too acid	Prefers soil slightly deficient in nitrogen
Purple Bean	Ordinary; pH 5.5 to 6.7	Prefers soil slightly deficient in nitrogen
Romano Bean	Ordinary; pH 5.5 to 6.7	Prefers soil slightly deficient in nitrogen
Scarlet Runner Bean	Ordinary; pH 5.5 to 6.7	Prefers soil slightly deficient in nitrogen
Soybean	Ordinary; pH 5.5 to 6.7	Prefers soil slightly deficient in nitrogen
Borage	Light, loamy	Average fertility

Vegetable	Soil Type	Nutrition
Asian Brassicas	Rich, loamy	High nitrogen requirements
Broccoli Raab	Ordinary; pH 6.5	Needs ample nitrogen. Responds well to some sifted compost
Domestic Burdock	Ordinary; deeply cultivated	Very average fertility
Salad Burnet	Ordinary, dry; fairly neutral	Prefers generally deficient soil
Nopal and Prickly Pear Cacti	Ordinary, well-drained	
Cardoon	Moist, loamy; not too acid	High nitrogen requirements
Purple Cauliflower/ Purple Broccoli	Ordinary; pH 6.0 to 6.8	High nitrogen requirements
Celeriac	Light, loamy; slightly acid	High potassium requirements
Celtuce	Ordinary; slightly acid	
Chayote	Very moist, well-drained	High nitrogen requirements
Chicory	Ordinary; slightly acid	
Collards	Ordinary; pH 5.5 to 6.5 is best	High nitrogen requirements
Comfrey	Any well-composted	
Coriander	Loamy, well-worked	Ordinary, unfertilized soil yields best flavor
Corn Salad	Ordinary	High nitrogen requirements
Cowpea	Prefers light soil; slight to moderate acidity	Prefers soil slightly deficient in nitrogen
Winter and Garden Cresses	Ordinary, cool	Moderate requirements of all major nutrients
Watercress	Moist, sandy	Moderate requirements of all major nutrients
Chinese Cucumber	Loam, rich in humus; pH less than 6.5	

Vegetable	Soil Type	Nutrition
Dandelion	Ordinary, prefers light	Responds well to added nitrogen and phosphorus
Daylily	Ordinary	
Japanese Eggplant	Moist, sandy, warm; neutral to slightly acid	High phosphorus and potassium requirements
Florence Fennel	Moist, cool; neutral	Moderate requirement of nitrogen
Garlic	Ordinary, moist; pH 5.8 to 6.5	
Elephant Garlic	Ordinary, moist; pH 5.8 to 6.5	
Good King Henry	Prefers rich, well-drained, but will grow in most soils	
Ground Cherry	Ordinary, well-prepared; slightly acid; tolerates high acidity	
Horseradish	Deep, moist loam with underlying clay	High amounts of potassium and humus required
Jerusalem Artichoke	Ordinary to poor	Moderate amounts of potassium and phosphorous required
Jicama	Light, rich, deeply cultivated	High potash requirements; likes wood ashes and well-rotted manure
Kale	Ordinary; pH 5.5 to 6.7	High nitrogen and calcium requirements
Flowering Kale	Ordinary; pH 5.5 to 6.7	High nitrogen and calcium requirements
Kohlrabi	Moist, light; pH 5.5 to 6.7	High nitrogen requirements
Leek	Ordinary, well-drained; pH 5.8 to 6.7	
Luffa	Ordinary, light; slightly acid	Responds well to manuring
Martynia	Moist, very loamy; pH 6.0 to 8.0	High requirements of all nutrients, especially nitrogen
Mustard	Ordinary, prefers light or sandy loam	Average fertility

Vegetable	Soil Type	Nutrition
Mustard Spinach	Ordinary, prefers light or sandy loam	Average fertility
Nasturtium	Light, poor soil	Prefers soil slightly low in all nutrients
Okra	Ordinary; pH 6.0 to 8.0	
Egyptian Onion	Ordinary; well-drained; pH 5.8 to 6.5	
Welsh Onion	Ordinary, well-drained; pH 5.8 to 6.5	
Orach	Ordinary, well-drained; tolerates saline, alkaline soils.	High nitrogen requirements
Hamburg Parsley	Sandy, light	High potassium requirements
Asparagus Pea	Rich, light loam	High nitrogen requirements
Sugar Pea	Ordinary, light; slightly acid to neutral	Prefers soil slightly low in nitrogen
Peanut	Very light; pH 5.0 to 6.0	
Popcorn	Ordinary, well-drained	High requirements of all nutrients, especially nitrogen
Sweet Potato	Ordinary; pH 5.2 to 6.7	High requirements of all nutrients, especially nitrogen
Lady Godiva Pumpkin	Light; slightly acid	Prefers soil slightly low in nitrogen
Purslane	Ordinary, even poor; tolerates acid soils	
Winter Radishes	Ordinary, light	
Rhubarb	Ordinary, prefers light; slightly acid	High potassium requirements
Rocket	Ordinary, moist	
Salsify	Light loam, deeply cultivated; neutral or slightly alkaline	High potassium requirements
Scorzonera	Ordinary, prefers light, deeply cultivated; neutral or slightly alkaline	High potassium requirements

Vegetable	Soil Type	Nutrition
Sea Kale	Ordinary	
Shallot	Ordinary, prefers light; pH 6.0 to 8.0	
Shungiku	Ordinary	
Skirret	Ordinary, well-drained, deeply cultivated; slightly acid	High potassium requirements
Sorrel	Ordinary	Reponds well to nitrogen fertilizer
New Zealand Spinach	Ordinary, deeply cultivated; pH 6.5 to 7.0	Fairly high nitrogen requirements
Spaghetti Squash	Ordinary, prefers light; slightly acid	
Tomatillo	Ordinary	Average fertility

Appendix 4

Optimum Climate Conditions

Vegetable	Preferred Climate	Hardiness	Location
Amaranth	Prefers hot days and cool nights	Resists some cold	Full sun
Bamboo	Moist, warm	Resists some cold	Partial shade
Basella	Warm, moist	Tender	Full to moderate sun
Asparagus Bean	Hot weather, can tolerate some drought	Very tender	Full sun
Fava Bean	Any	Resists some cold	Full sun
Horticultural Bean	Cool	Fairly hardy	Full sun
Purple Bean	Warm	Tender	Full sun
Romano Bean	Warm	Tender	Full sun
Scarlet Runner	Warm, hot	Tender	Full sun
Soybean	Warm; special varieties for north	Tender	Full sun
Borage	Warm; greenhouse	Fairly hardy	Prefers full sun, but can stand some shade
Asian Brassicas	Cool, moist	Resist some cold	Full sun
Broccoli Raab	Any; cool	Very hardy	Full sun
Domestic Burdock	Any	Fairly hardy	Full to partial sun

Vegetable	Preferred Climate	Hardiness	Location
Salad Burnet	Any; warm	Hardy	Full sun
Nopal and Prickly Pear Cacti	Warm, dry	Hardy	Full sun
Cardoon	Warm	Tender	Full sun
Purple Cauliflower/ Purple Broccoli	Cool weather	Tolerates some heat	Full sun but can stand some shade.
Celeriac	Cool; 60 to 65°F. mean during growing season	Very hardy; resists frosts, some heat	Full to partial sun
Celtuce	Cool	Resists some frost	Full to partial sun
Chayote	Warm; greenhouse	Very tender	Full sun
Chicory	Cool, moist	Fairly hardy	Full to partial sun
Collards	Any	Withstands considerable frosts	Any
Comfrey	Moist; can survive drought	Resists some cold	Moderate sun; tolerates full sun and partial shade
Coriander	Cool	Resists light frost; bolts in warm temperatures	Full sun
Corn Salad	Cool	Cannot tolerate hot weather; resists light frosts and winters over where protected	Partial to full sun
Cowpea	Warm, tropical or subtropical	Very tender	Full sun
Daylily	Any	Very hardy	Sun or shade
Winter and Garden Cresses	Cool	Cannot tolerate very hot weather	Full sun
Watercress	Cool	Hardy	Partial sun or shade
Chinese Cucumber	Warm, moist	Resists some cold	Full sun

402

Vegetable	Preferred Climate	Hardiness	Location
Dandelion	Anywhere except Deep South	Very hardy	Prefers sun; tolerates much shade
Japanese Eggplant	Hot	Tender	Full to moderate sun
Florence Fennel	Cool, moist	Bolts in hot weather	Full sun
Garlic	Cool; will grow anywhere	Very hardy	Full to moderate sun
Elephant Garlic	Cool, will grow anywhere	Hardy	Full to moderate sun
Good King Henry	Any	Hardy	Any, even full shade
Ground Cherry	Warm; greenhouse	Tender	Full sun
Horseradish	Cool	Very hardy	Full to moderate sun
Jerusalem Artichoke	Any; prefers cool	Very hardy	Any, sunny
Jicama	Warm; nine warm months for large tubers	Tender	Moderate to full sun
Kale	Cool	Most cold-hardy of its family	Full to moderate or even partial sun
Flowering Kale	Warm; greenhouse	Fairly tender	Full sun
Kohlrabi	Cool	Tolerates some freezing	Full to moderate sun
Leek	Cool; any	Very hardy	Moderate sun
Luffa	Hot	Tender	Full sun
Martynia	Warm	Tender	Full sun
Mustard	Cool	Hardy; bolts in warm weather	Full to moderate sun
Mustard Spinach	Cool	Very hardy; tolerates some warm weather	Full to moderate sun
Nasturtium	Any; warm; greenhouse	Resists some cool and very hot temperatures but is not hardy	Full sun
Nettle	Any	Hardy	Any

Vegetable	Preferred Climate	Hardiness	Location
Okra	Warm	Some varieties resist cool temperatures; tender	Full sun
Egyptian Onion	Any	Hardy	Any
Welsh Onion	Any; cool	Very hardy	Any
Orach	Cool	Hardy; withstands heat	Prefers shade or moderate sun
Hamburg Parsley	Cool	Very hardy	Any
Asparagus Pea	Warm	Tender	Full sun
Sugar Pea	Cool, moist	Fairly hardy	Full sun
Peanut	Warm, but tolerates cooler temperatures; greenhouse	Tender	Full sun
Popcorn	Warm, moist	Not hardy; varieties resist cold	Full sun
Sweet Potato	Warm; 4 to 5 frost-free months; greenhouse	Tender	Full sun
Lady Godiva Pumpkin	Warm	Withstands light frosts	Full sun
Purslane	Any	Hardy	Any
Winter Radishes	Cool	Hardy	Full to moderate sun
Rhubarb	Cool	Hardy	Full to moderate sun
Rocket	Cool	Hardy	Full sun
Salsify	Any; prefers cool	Tolerates frosts; winters over when protected with mulch	Moderate to full sun
Scorzonera	Any	Tolerates some frost	Moderate to full sun
Sea Kale	Any	Resists some light frosts but must be heavily mulched to endure cold winters	Full sun

Vegetable	Preferred Climate	Hardiness	Location
Shallot	Any	Resists some frosts	Full to moderate sun
Shungiku	Cool	Fairly hardy	Full to partial sun
Skirret	Cool	Hardy	Full to moderate sun
Sorrel	Any, moist	Hardy	Full to moderate sun
New Zealand Spinach	Warm, but not hot; moist	Very tender	Full to moderate sun
Spaghetti Squash	Any	Tender	Full sun
Tomatillo	Hot, tropical; greenhouse	Tender	Full sun

Appendix 5

Making the Best Use of Your Garden

Vegetable	*Spacing Between Plants*	*Plant Size*	*Harvest*
Amaranth	10 inches apart	Vegetable amaranth: 6 to 10 inches tall if kept cut Grain crop: 6 to 8 feet tall	Greens harvested just 3 or 4 weeks after sowing 120 days required to produce mature grain crop
Bamboo	6 to 15 feet apart	Height of 20 feet, but some smaller types are available.	After first year, in spring and throughout the season
Basella	1 foot apart	Large climber	Continuously until frost
Asparagus Bean	2 inches deep, 3 seeds per pole or in rows 8 inches apart	Climbs 7 to 10 feet	Throughout the summer
Fava Bean	2 inches deep, 3 to 4 inches apart in rows 20 inches apart	Climber usually 3 to 5 feet tall, but some dwarf varieties available	Late spring in most regions; winter in gulf and coastal areas
Horticultural Bean	1 inch deep, 2 inches apart (thin plants to 1 foot apart)	Bush, 1½ to 2 feet in height	60 days for green shell beans, less for snap beans; 90 days for dried beans
Purple Bean	1 inch deep in rows 10 to 12 inches apart	Compact plants, 1 to 2 feet in height	50 to 60 days after seeding

Vegetable	Spacing Between Plants	Plant Size	Harvest
Romano Bean	1 inch deep, 3 to 4 inches apart in rows 18 inches apart	Climber; grows tall but not too thick	55 to 70 days after seeding
Scarlet Runner Bean	1 inch deep, 8 inches apart in rows 18 to 24 inches apart	Climber; forms an excellent screen when trellised	Continuously from mid-summer through early fall for snap beans; 120 days for full maturity
Soybean	1 to 1½ inches deep, 4 inches apart; less for low-growing varieties	About 2 feet in height	Green beans harvested from mid through late summer; 90 to 115 days to full maturity
Borage	2 feet apart in rows 2 feet apart	2 to 4 feet in height	Early summer; succession plantings can be harvested throughout the season
Broccoli Raab	2 to 5 inches apart	12 to 18 inches tall	Early spring
Domestic Burdock	1½ feet apart in rows about 2 feet apart	Very large leaves that tend to shade nearby plants	Late fall or early spring; 3 to 5 months after sowing
Salad Burnet	⅛ inch deep; 8 to 10 inches apart in rows about 12 inches apart	Low and sprawling; 1 foot high at flowering time	Year-round
Asian Brassicas Chinese Cabbage	1 foot apart	Heads 10 inches high, 6 to 7 inches in diameter	70 to 80 days after planting
Michihli	10 inches apart	About 18 inches tall	Mid to late summer; about 80 days after seeding
Bok Choy	1 foot apart	About 1 foot high	50 to 60 days after seeding for table use; 30 days for salad greens
Nopal and Prickly Pear Cacti	Varies with variety	Variable, from prostrate to 6 or 10 feet	Small nopalitos: anytime; pear: midsummer through late fall

Vegetable	Spacing Between Plants	Plant Size	Harvest
Cardoon	3 feet apart in rows	Very large plants	After stalks are 1 to 2 feet high and throughout the season
Purple Cauliflower/ Purple Broccoli	18 inches apart	6- to 8-inch heads	Late summer through early fall and late fall; 90 to 95 days after seeding
Celeriac	Transplants set out 6 inches apart in rows 12 inches apart	1½ feet tall and about 1 foot wide	Throughout the fall and winter, if mulched; about 120 days after plants have been set outside
Celtuce	1 foot apart in rows 18 inches apart	Height of celery	Throughout the fall; 45 to 100 days after seeding
Chayote	In hills 9 to 12 feet apart	Climbing, 30 to 100 feet; requires a lot of space	Shoots: late spring through early summer; fruits: late summer; tubers: fall
Chicory	6 inches apart in rows 18 inches apart	Similar to lettuce; depends on variety	Leaves: early spring; heads: fall, about 90 days after planting; forcing: roots pulled in late fall after first frost but before long winter freeze sets in
Collards	18 to 24 inches apart in rows 3 feet apart	3 to 5 feet tall	Midsummer
Comfrey	12 inches apart in rows 12 inches apart	2 feet tall	Several times a year or throughout the season
Coriander	3 to 4 inches apart	12 to 30 inches tall	Early in spring as soon as mature leaves begin to appear
Corn Salad	4 inches apart in rows 1 foot apart	1½ feet tall	Fall, spring, and in some regions, throughout the winter
Cowpea	3 inches apart in rows 3 feet apart	Climbing and bush varieties available	Varies; 8 to 10 weeks after planting, continuing for 2 months

Vegetable	Spacing Between Plants	Plant Size	Harvest
Winter and Garden Cresses	2 to 3 inches apart in rows about 9 inches apart	10 to 30 inches tall, compact	Spring, early summer, and fall; 10 days to 2 weeks after sowing
Watercress	Plants spaced 6 inches apart	Trailing	Late spring through early winter; year-round in mild regions
Chinese Cucumber	1 foot apart in hills or rows	Tall climber	Throughout the season beginning about 50 days after sowing
Dandelion	18 to 24 inches apart in rows 10 inches apart	About 2 feet wide	Best in early spring; 6 weeks to 95 days after planting
Daylily	Permanent beds	About 3 feet tall; will spread if left unchecked	Young leaves: early spring; flowers: midsummer; tubers: late fall or winter
Japanese Eggplant	3 feet apart in rows 3 feet apart	2 feet tall	Mid through late summer; 70 to 90 days after sowing
Florence Fennel	6 to 8 inches apart in rows 18 inches apart	1 to 2 feet tall	Summer in cool regions; autumn, winter, or spring in milder areas. 80 to 110 days after seeding
Garlic	About 6 inches apart	About 10 inches tall	Green leaves harvested year-round; bulbs in fall or winter, after tops have fallen
Elephant Garlic	12 inches apart in rows 2 to 3 feet apart	30 to 36 inches tall	3 to 4 months after planting, when tops have fallen
Good King Henry	Rows 18 inches apart	2 feet tall	Best in late spring through early summer; can be eaten throughout the season
Ground Cherry	3 feet apart in rows 3 feet apart	Spreading; about 1 foot tall	Once a week from mid to late summer beginning 75 to 90 days after seeding
Horseradish	Root slips planted 1 foot apart in rows 2 feet apart	Compact	Fall or early spring

Vegetable	Spacing Between Plants	Plant Size	Harvest
Jerusalem Artichoke	Hills 1 foot apart or in rows 3 to 4 feet apart	6 to 12 feet high	Fall, when the plants die down; also throughout the winter and in early spring
Jicama	6 to 8 inches apart in rows 1 foot apart	Very large vines, 20 to 25 feet long	Late summer; in warm regions, throughout the year
Kale	About 1 foot apart in rows 18 inches apart	Varies according to variety; 1 to 3 feet tall	Spring or fall; throughout the winter and early spring in the South; 55 to 65 days after planting
Flowering Kale	8 inches apart in rows 2 feet apart	Low and compact	Late summer through fall; 80 to 90 days after seeding
Kohlrabi	6 inches apart in rows 15 to 18 inches apart	About a foot high	Early summer through early fall; 55 to 75 days after seeding
Leek	6 to 8 inches apart in rows 12 inches apart	About 1 foot tall	Late summer and throughout the fall, winter, and early spring. 150 to 190 days after seeding
Luffa	Hills 6 feet apart	Climbing vines, should be trained	Fall, before the first frost
Martynia	4 feet apart in rows 3 to 5 feet apart	Very wide and spreading; 2 feet tall	2 months after planting
Mustard	1 to 2 inches apart in rows 12 to 15 inches apart	2 to 6 feet high, depending on variety	Late fall in most regions: 40 to 50 days after seeding
Mustard Spinach	1 to 2 inches apart in rows 12 to 15 inches apart	Similar to curly mustard varieties	Throughout the season, beginning 25 to 30 days after seeding
Nasturtium	In rows 4 feet apart	Climbing or dwarf, compact	Leaves: throughout the season; flowers: midsummer through fall; seeds: late summer through early fall
Nettle	In beds, a few inches apart.	Up to 7 or 8 feet high	Early spring through early summer, while the tops are 6 inches tall

Vegetable	Spacing Between Plants	Plant Size	Harvest
Okra	8 to 12 inches apart in rows 4 to 6 feet apart; dwarf types in rows 2 to 4 feet apart	4 to 8 feet tall; some dwarf varieties only 2 to 4 feet tall	55 to 65 days after seeding; 2 harvests during the season
Egyptian Onion	Sets planted 5 to 6 inches apart in rows about 10 inches apart	About 18 inches tall	Continuously
Welsh Onion	4 inches apart	About 20 inches tall	Early spring and winter; 60 days after planting
Orach	2 feet apart	4 to 10 feet tall, depending on variety; large leaves	After plants are well established, young tips harvested continuously through summer
Hamburg Parsley	6 to 8 inches apart	Compact	Fall and spring; about 4 months after planting
Asparagus Pea	12 inches apart in rows 15 inches apart	Climber, but doesn't take much space	When pods are 1 inch long; produces for 10 weeks
Sugar Pea	Dwarf: 3 inches apart in rows 18 to 24 inches apart; climbers: 2 inches apart in double rows	Tall climber; some dwarf varieties 18 inches tall	50 to 60 days after seeding; spring in most regions
Peanut	3 feet apart	12 to 24 inches tall	Fall, 100 to 120 days after seeding
Popcorn	6 to 12 inches apart	6 to 10 feet tall, depending on variety	100 to 120 days
Sweet Potato	18 inches apart in rows 3½ feet apart	Twining and trailing	Late summer to early fall
Lady Godiva Pumpkin	In hills 3 to 5 feet apart	Sprawling	110 to 115 days after planting
Purslane	In rows or beds	Low, sprawling	Continuously from early summer through frost

Vegetable	Spacing Between Plants	Plant Size	Harvest
Winter Radishes	Thinly in rows 15 to 24 inches apart	Varies according to variety	Fall, early winter harvest, about 50 to 60 days after seeding
Rhubarb	Plants 2 to 4 feet apart in rows 4 to 6 feet apart	Very large, bushy. Up to 3 or 4 feet high	3 years from seed or 1 to 2 from crowns. Harvest stalks in spring or early summer
Rocket	6 to 9 inches apart in rows 16 inches apart	Erect, to 2½ feet tall	Early spring, about 5 weeks after sowing and continuously if desired
Salsify	About 5 inches apart in rows 2 feet apart	If left as a perennial, will grow up to 4 feet tall; otherwise compact	4 months after planting and throughout the winter
Scorzonera	About 5 inches apart in rows 2 feet apart	Compact, like salsify	Early autumn throughout the winter
Sea Kale	Plants should be 2 to 3 feet apart in rows 3 feet apart	Coarse, broad; 2 to 3 feet high	2 years after planting from seed; 1 year after planting from cuttings. Harvest shoots as they appear in spring or gather for forcing in winter
Shallot	4 inches apart	2 feet tall	Top growth can be cut throughout the year; bulbs in the fall
Shungiku	1 to 2 inches apart in rows 18 inches apart	Plants grow large and must be cut frequently to about 12 inches	Early summer and spring, about 6 or 7 weeks after sowing
Skirret	9 to 12 inches apart in rows 18 inches apart	2 feet tall but delicate and will not shade nearby plants	October; 5 to 6 months after planting
Sorrel	4 inches apart	12 to 36 inches tall; will clump	Especially choice in spring; otherwise throughout the season except during extreme cold

Vegetable	Spacing Between Plants	Plant Size	Harvest
New Zealand Spinach	1 to 2 inches apart	Fairly large plant after 2 months	All of the leaves several times during the season or 3- to 4-inch tips continuously
Spaghetti Squash	3 plants per hill; hills 6 to 8 feet apart	Vining, very dense	Late summer to early fall, 90 to 110 days after planting
Tomatillo	1 foot apart in rows 1 foot apart	Spreading	Fall

Appendix 6

By food value, we mean simply why each vegetable is valuable as a food, and why gardeners like to grow it. This chart lists the special characteristics which make each vegetable excel in the kitchen and garden.

Vegetable	*Value as a Food and Garden Crop*
Amaranth	Leaves contain substantial amounts of vitamins A and C; dried grain is 20 percent protein; tolerant of hot weather; long-producing.
Bamboo	Crisp, chewy texture irreplaceable in many oriental dishes; graceful trees attractive for landscaping; withstands much cold.
Basella	Leaves contain lots of vitamin A and are extremely rich in vitamin C; delicate flavor; thrives in hot weather which causes spinach to bolt.
Asparagus Bean	Excellent, distinctive flavor; good producers; thrives in hot regions where snap beans fail.
Fava Bean	Good flavor; excellent early shelling bean; freezes well; cold-resistant; heavy producer.
Horticultural Bean	Superior freezing and canning qualities; good flavor; easy to shell; versatile in the kitchen.

414

Vegetable	Value as a Food and Garden Crop
Purple Bean	Stringless and full-flavored; best variety for freezing since color changes to green when blanching is complete; heavy producer.
Romano Bean	Excellent flavor; especially good for freezing; heavy producer; bears continuously until frost.
Scarlet Runner Bean	Hearty flavor; rich in B vitamins, iron, and calcium; versatile in the kitchen; foliage and flowers make lovely ornamental screen.
Soybean	Outstanding nutritional value: best vegetable source of protein, lots of B vitamins and calcium, vitamin E; versatile in the kitchen.
Borage	Distinctive flavor; pretty flowers good for attracting bees.
Broccoli Raab	Pleasant flavor; among the earliest garden crops; needs little care.
Domestic Burdock	Distinctive flavor; medicinal value as well; stores well.
Salad Burnet	Winter-hardy; distinctive flavor.
Asian Brassicas: Chinese Cabbage and Michihli	Fine flavor and texture; some varieties store quite well.
Bok Choy	Distinctive flavor and uses; substantial amounts of vitamins A and C; good warm-weather greens.
Nopal and Prickly Pear Cacti	Special flavor; fruits good thirst-quenchers; has vitamin C.
Cardoon	Unique flavor; stores well—fine winter vegetable; versatile in the kitchen.
Purple Cauliflower/ Purple Broccoli	Excellent flavor; stands up to heat better than white cauliflower or green broccoli; fairly high in vitamin C; freezes very well.

Vegetable	Value as a Food and Garden Crop
Celeriac	Excellent flavor; stores quite well; requires no blanching; yields heavily and is frost-hardy.
Celtuce	Distinctive flavor; high in vitamin C and many minerals, including iron; versatile in the kitchen—can be used as two separate vegetables.
Chayote	Extremely versatile in the kitchen; young shoots, young fruits, older fruits, and tubers all used differently; flourishes in hot weather.
Chicory	Distinctive flavor; low in calories and high in vitamin A; versatile in the kitchen—all parts edible.
Collards	Outstanding nutritional value—rich in vitamins A, B, and C and many minerals; good flavor; stands a lot of frost.
Comfrey	Excellent nutritional value—rich in vitamins A, C, E, and perhaps B_{12}; also contains calcium, phosphorus, potassium, and as much protein as alfalfa; medicinal value as well.
Coriander	Distinctive flavor absolutely necessary to certain Mexican and oriental dishes; high in vitamins A and C, phosphorus and iron.
Corn Salad	Outstanding flavor and texture; very frost-hardy; fine winter salad vegetable.
Cowpea	Contains protein; meaty and substantial flavor; versatile in the kitchen; freezes well.
All Cresses	Distinctive flavor; very high in vitamin C.
Chinese Cucumber	Very crisp; few and small seeds; does not cause indigestion; good for either slicing or pickling; hardier than domestic cucumbers.
Dandelion	Rich in vitamin A and minerals; distinctive flavor; versatile in the kitchen—leaves, roots, and flowers all used; available year-round in many regions.

Vegetable	Value as a Food and Garden Crop
Daylily	Extremely versatile in the kitchen—young leaves, flower buds, open and wilted flowers, and tubers all used; lovely, easy-to-grow ornamental as well.
Japanese Eggplant	Superior flavor; thin skin makes it easier to cook with than other varieties; earlier than domestic varieties.
Florence Fennel	Excellent, distinctive flavor; extremely dependable cool-season vegetable.
Garlic	Distinctive flavor; versatile in the kitchen; medicinal value as well.
Elephant Garlic	Excellent flavor—milder and sweeter than garlic; can be used as cooked vegetable as well as seasoning; high yielder.
Good King Henry	Versatile in the kitchen; hardy and easy to grow.
Ground Cherry	Excellent flavor; cheaper and better-tasting than citron in fruit cakes; very productive.
Horseradish	Unmistakable flavor; excellent winter storage qualities; homegrown surpasses commercial prepared product.
Jerusalem Artichoke	Contains inulin and levulose which are valuable to diabetics; less starchy than potatoes; stores well throughout the winter—completely hardy; high yielder.
Jicama	Unique flavor and texture; easier to grow than water chestnuts.
Kale	Outstanding nutritional value—among the best of any vegetable; rich in vitamins A, B, and C, calcium and iron; good flavor; stands up well to frost.
Flowering Kale	Good flavor; heads are ornamental as well as edible.
Kohlrabi	Unique flavor; good nutritional value—contains vitamins A and B.

Vegetable	Value as a Food and Garden Crop
Leek	Outstanding flavor; good source of vitamin B; very hardy.
Luffa	Young fruits good source of vitamin C and fiber; older fruits make excellent sponges.
Martynia	Young seedpods good for pickling and soups; easy-to-grow ornamental.
Mustard	Distinctive flavor; good nutritional value—substantial amounts of vitamins A, B, and C, calcium and iron.
Mustard Spinach	Distinctive flavor, milder than mustard; also good source of vitamins A and C, calcium and iron; very quick-growing.
Nasturtium	Distinctive flavor; high in vitamin C; versatile in the kitchen.
Nettle	Good nutritional value—substantial amounts of vitamins A and C, iron, and protein; pleasant flavor; medicinal value and many other uses.
Okra	Distinctive flavor; excellent thickening agent for stews and soups.
Egyptian Onion	Strong flavor, good for pickling; produces continuously.
Welsh Onion	Delicate flavor; hardy perennial crop.
Orach	Good flavor; large, tender leaves; hardy and fast-growing.
Hamburg Parsley	Substantial amounts of vitamins A and C; low in calories; distinctive flavor; versatile in the kitchen.
Asparagus Pea	Unique flavor; no waste; heavy producer.
Sugar Pea	Excellent flavor; no waste.
Peanut	Good source of protein; high potassium and low sodium content makes it good for low-sodium diets.

Vegetable	Value as a Food and Garden Crop
Popcorn	Homegrown has better flavor than most store-bought types.
Sweet Potato	Very rich in vitamin A and carotene (which our bodies convert to vitamin A); distinctive flavor. Since they don't take to shipping and handling, best-quality sweets are grown at home.
Lady Godiva Pumpkin	Seeds are high in protein and iron, and require no shelling.
Purslane	Good flavor; contains vitamins B and C; medicinal uses; early, easy-to-grow greens.
Winter Radishes	Milder flavor than spring/summer radishes; more versatile in the kitchen; store well over the winter.
Rhubarb	Distinctive flavor; best when picked fresh from the garden.
Rocket	Distinctive flavor; early, hardy crop.
Salsify	Unique flavor; very hardy late crop.
Scorzonera	Flavor distinctive like salsify; roots contain inulin, good source of carbohydrate for diabetics; very hardy.
Sea Kale	Unique, succulent flavor; hardy perennial.
Shallot	Outstanding flavor; heavy producer.
Shungiku	Distinctive flavor, especially good in oriental dishes.
Skirret	Distinctive flavor, attractive in the garden.
Sorrel	Extremely rich in vitamin A, also high in vitamin C and iron; distinctive flavor.
New Zealand Spinach	Good flavor; thrives in hot weather that causes spinach to bolt.
Spaghetti Squash	Low in calories, good substitute for spaghetti; stores well; early producer.
Tomatillo	Unique flavor an absolute must in certain Mexican and oriental dishes.

Appendix 7

Vegetable	*Flavor*	*Substitutions*	*Parts Used and Their Use*
Amaranth	pleasant; unlike any other green vegetable	leaves for spinach; seeds for other grains	mature stems–H
			leaves–C, F, 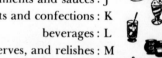, H, I
			young shoots–H
			seeds–E, F, K, L
Bamboo	delicate, somewhat sweet, depending on variety; reminiscent of young field corn		young shoots–B, C, H, I, M
			leaves–
			poles–R

*KEY TO USES

appetizers, hors d'oeuvres, and snacks: A
salads: B
soups: C
sandwiches: D
breads and cakes: E
cereals and pasta dishes: F
egg dishes: G
vegetable side dishes: H
main dish casseroles and combinations: I

condiments and sauces: J
desserts and confections: K
beverages: L
pickles, preserves, and relishes: M
freeze (can be frozen): N
can (can be canned): O
medicinal uses: P
livestock forage: Q
other household uses: R
ornamental: S

Vegetable	Flavor	Substitutions	Parts Used and Their Use
Basella	delicate, mild	spinach	leaves–C, H
Asparagus Bean	subtle, zippy, nutty, pea/bean taste; like asparagus to some	snap beans	young beans–[icon], H, I young leaves and shoots–H
Fava Bean	sweeter than limas	green limas; also snap beans and dried beans	young beans–H developed beans–A, H, [icon] mature beans–A, M young leaves–H
Horticultural Bean	similar to lima beans, but heartier	green limas; also snap beans and dried beans	young beans–A, B, H developed beans–H, N mature beans–[icon]
Purple Beans	like snap beans	snap beans	young beans–B, H, N, O
Romano Beans	complex, delicate; like snap beans, but different	snap beans; also dried beans	young beans–H developed beans–[icon], H, I
Scarlet Runner Bean	like snap beans, but rather nutty; hearty	snap beans; also green limas and dried beans	young beans–B, H, N, O developed beans–[icon], H, I mature beans–C, H
Soybeans	buttery, unique	green shell beans and dried beans in some recipes	immature beans–A, B, C, H, I, [icon], O mature beans–C, M, I, L
Borage	refreshing, mild, cucumber-like		flowers–[icon] young leaves–B, H, L, P, Q

Vegetable	Flavor	Substititions	Parts Used and Their Use
Asian Brassicas: Michihli	similar to Chinese cabbage, but more pungent		leaves–C, H, I
Chinese Cabbage	similar to cabbage, but finer in taste		leaves– , C, H, I, M
Bok Choy	unique; tender, sweet, crisp, milder than either of the above		leaves and stalks–B, C, H, I
Broccoli Raab	succulent, with taste of cabbage family; somewhat like asparagus	asparagus	young shoots–H
Domestic Burdock	refreshing, sweetish, unusually aromatic	carrot; parsnips	roots–C, H, I, , M
Salad Burnet	distinctive, cucumber-like	other distinctive greens	leaves–B, C, D, L, P
Nopal and Prickly Pear Cacti	pads: bland; similar to green pepper; fruits: like melon; very refreshing	pads for okra; fruits for melon	pads–G, H, I fruits– K, L
Cardoon	slightly bitter; resembles artichoke		leaf stalks–B, C, H, M, O seedheads–
Purple Cauliflower/ Purple Broccoli	milder than broccoli, richer than cauliflower	broccoli, cauliflower	immature flower heads–A, B, C, G, H, stems–B, H
Celeriac	like celery, but milder, mellower	celery in cooked dishes, turnips	bulbs–B, C, H leaves, sparingly–B, C
Celtuce	stalk: cool, similar to cucumber or summer squash; leaves: bitter	stalks for celery, leaves for lettuce	stalks–A, B, C, , I young leaves–B

Vegetable	Flavor	Substitutions	Parts Used and Their Use
Chayote	fruits: variable; refreshing, similar to summer squash; sweeter as fruits age; tubers: starchy; young shoots: subtle flavor somewhere between celery and cabbage	shoots for asparagus, tubers for potatoes; young fruits for cucumbers, older fruits for eggplant, fully mature fruits for potatoes	shoots–B, H tubers–H, K fruits–B, C, [image], H, I, K, M
Chicory	leaves and roots: bitter; forced shoots: mild, delicate	leaves for lettuce; roots for parsnips, coffee	leaves–B, H roots–H, [image] forced heads–B, C, H
Collards	strong; like the cabbage family		leaves–H, I [image]
Comfrey	pleasant, faint taste of cucumber, less pronounced than borage	other distinctive greens	young leaves–B, C, H, K, L, N, [image], Q
Coriander	pungent, aromatic, distinctive		leaves–G, H, I, J
Corn Salad	delicate; melt-in-your-mouth	lettuce or other bland greens	leaves–[image], H
Cowpea	meaty, hearty	dried beans; also snap beans and green shell beans	young beans–B, H developed beans–H mature beans–C, [image]

Cresses:
    Winter and
    Garden Cresses
	tangy, pungent, spicy-hot		leaves–B, C, [image], H
			
			seed stalks–S

Watercress	zingy and pungent but no afterburn		leaves–B, C, D, [image], H, I

Vegetable	Flavor	Substitutions	Parts Used and Their Use
Chinese Cucumber	like cucumber, especially crisp	regular cucumbers	fruits–A, , C, D, H, M
Dandelion	leaves and roots: bitter; forced leaves: mild	root for coffee	leaves–B, D, H flowers– G, H, roots–H, L, P forced shoots–B, H
Daylily	leaves: mild; flowers delicate; tubers: nutty	tubers for potatoes	leaves–H flowers–B, H, tubers–A, B, H
Japanese Eggplant	like eggplant, but more delicate and distinctive	regular eggplant	fruits–A, C, H, , M
Florence Fennel	sweet, mildly anise	bulbs for celery	bulbs–B, H leaves–B, I, L, P seeds–K, P
Garlic	pungent, distinctive	tops for chives	bulbs– C, , H, I, J, P tops–same
Elephant Garlic	like garlic but milder, sweeter	bulbs for true garlic	bulbs–B, C, F, H, I,
Good King Henry	leaves: bitter, like spinach; stalks: like asparagus	leaves for spinach; stalks for asparagus	leaves–H stalks–H
Ground Cherry	sweet-tart, slightly acidic	candied fruits for citron	fruits–B, E, K,

Vegetable	Flavor	Substitutions	Parts Used and Their Use
Horseradish	quite pungent; forced leaves: sweet and mild		roots–B, H, J, M, P leaves– 🥗, H
Jerusalem Artichoke	nutty, sweet	potatoes	tubers–B, C, 🔥, I, M
Jicama	unique; crispy-sweet	water chestnuts	tubers– 🥔, B, H, I
Kale	like collards but milder; rich with a hint of the cabbage family		leaves–B, C, 📦, H, Q
Flowering Kale	like kale	cabbage	leaves–B, H, I, 🌸
Kohlrabi	cross between turnip and apple		bulbs–A, B, H, 🥣, N
Leek	oniony, but very mild and smooth; richer than onion	onion	thick lower part of stems (not really a bulb)– ∞ B, C, G, H
Luffa	mild	young fruits for summer squash, okra, or cucumbers	young fruits–B, C, A mature fruits–R
Martynia	distinctive; similar to okra	okra or cucumbers	young seedpods–C, M dried pods–S
Mustard	pungent, mild to hot depending on variety and growing conditions	spinach or chard	leaves–B, D, H, Q seeds– 🍳, P
Mustard Spinach	mild and savory	same as above	leaves–B, D, H, Q

Vegetable	Flavor	Substitutions	Parts Used and Their Use
Nasturtium	leaves: tangy like watercress; flowers: peppery and hot, like radishes; seeds: pungent	pickled seeds for capers	young leaves–B, D flowers–B, D, J, S seeds–
Nettle	salty, earthy	spinach	young leaves– C, H, L, P, Q,
Okra	distinctive		young pods–B, C, H, , M, N, O mature pods–S leaves–H ripe seeds–H, L, M
Egyptian Onion	stronger and richer than other onions	bulbs for onion; young stalks for green onion; stems for chives	bulbs–B, C, H, M young stalks–B, G,
Welsh Onion	delicate	green onion	stalks–B, C, H, I
Orach	mild; not acidy	spinach	leaves–H
Hamburg Parsley	like parsley-flavored celeriac	roots for parsnips; leaves for parsley	roots and leaves–, C, H
Asparagus Pea	quite unique; like combination of asparagus and peas		young pods–H
Sugar Pea	pea-like, but sweeter		young pods–B, H, I,
Peanut	earthy, nutty		seeds (nuts)–A, B, C, E, I, K
Popcorn	nutty	ground, for cornmeal	dried kernels– A, , S
Sweet Potato	mealy, sweet, rich and heavy	yams; pumpkin	tubers–A, B, C, H, K

Vegetable	Flavor	Substitutions	Parts Used and Their Use
Lady Godiva Pumpkin	seeds: nutty, almost earthy; immature fruit: similar to summer squash	immature fruit for summer squash or avocado	immature fruit–B, C, H seeds–
Purslane	tart, savory, slightly acid	other distinctive greens	leaves–B, C, H, P
Winter Radishes	milder than spring/summer radishes	turnips	roots–B, H
Rhubarb	tart		stalks–J, K, M
Rocket	sharp, spicy, pungent	other pungent greens	young leaves–B, , H older leaves–C
Salsify	sweet, oyster-like	roots for parsnips; shoots for asparagus	roots–B, H young shoots– young leaves–B, H
Scorzonera	oyster-like, more pronounced than salsify	roots for parsnips	roots–C, H
Sea Kale	delicate, nutty, slightly bitter	asparagus	stalks (shoots)–A, B, H
Shallot	light; subtler than onion or garlic, but stronger than leek	bulbs for onion; leaves for chives; young bulbs for green onion	all parts–B, G, H,
Shungiku	pungent and aromatic, like chrysanthemum smells	other pungent leaves	young leaves–B, C, H, I flowers–S
Skirret	sweet, unusual	parsnips	roots– , C, H
Sorrel	tart, tangy	spinach or other salad greens	leaves–B, C, H, J, P

Vegetable	Flavor	Substitutions	Parts Used and Their Use
New Zealand Spinach	like spinach but milder	spinach	young shoots–B leaves–H
Spaghetti Squash	sweet squash flavor	mature fruits for spaghetti; immature fruits for eggplant	mature fruits–B, 🦌, I immature fruits–H
Tomatillo	similar to ground cherry, but less sweet and not as pronounced; remind some of green tomatoes		fruits–A, B, 🍳, O

Bibliography

Angier, Bradford. *Gourmet Cooking for Free.* Harrisburg, Pa.: Stackpole Books, 1970.

Staff of the Liberty Hyde Bailey Hortorium. *Hortus Third.* N.Y.: MacMillan Publishing Co., Inc., 1976.

Bianchini, Francesco; Corbetta, Francesco; and Pistoia, Marilena. *The Complete Book of Fruits and Vegetables.* New York: Crown Publishers, Inc., 1975.

Brouk, B. *Plants Consumed by Man.* London: Academic Press, 1975.

Camp, Wendell H.; Boswell, Victor R.; and Magness, John R. *The World in Your Garden.* Washington, D.C.: National Geographic Society, 1957.

Cooper, Ed Wilfrid-Shewell. *The Complete Vegetable Grower.* London: Faber and Faber, 1965.

Day, Harvey. *About Rice and Lentils.* London: Thorsons Publishers Limited, 1970.

Eating Wild. Compiled by the Edible Wild Foods Committee, Ipswich River Wildlife Sanctuary. Massachusetts Audubon Society, 1971.

Edlin, H.L. *Plants and Man, The Story of Our Basic Food.* Garden City, N.Y.: The Natural History Press, 1969.

Encyclopedia Britannica. Chicago: Encyclopedia Britannica, Inc., 1973.

Genders, Roy. *The Complete Book of Vegetables and Herbs.* London: Ward Lock, 1972.

Gibbons, Euell. *Stalking the Healthful Herbs.* New York: David McCay Co., Inc., 1966.

Graber, Kay. *Nebraska Pioneer Cookbook.* Lincoln, Nebraska: University of Nebraska Press, 1974.

Harrison, S.G.; Masefield, G.B.; Nicholson, B.E.; and Wallis, M. *The Oxford Book of Food Plants.* London: Oxford University Press, 1969.

Hawkes, Alex D. *A World of Vegetable Cookery.* New York: Simon and Schuster, 1968.

Hills, Lawrence D. *Russian Comfrey.* London: Faber and Faber, 1953.

Hyams, Edward. *Plants in the Service of Man.* Philadelphia and New York: J.B. Lippincott Co., 1971.

Jones, Dorothea Van Gundy. *The Soybean Cookbook.* New York: Arc Books, Inc. by arrangement with the Devin-Adair Co., 1963.

Kadans, Joseph M. *Encyclopedia of Fruits, Vegetables, Nuts and Seeds.* New York: Parker Publishing Co., Inc., 1973.

Keys, John D. *Chinese Herbs, Their Botany, Chemistry, and Pharcodynamics.* Rutland, Vermont: Charles E. Tuttle Co., 1976.

Kraft, Ken and Pat. *The Best of American Gardening.* New York: Walker and Company, 1975.

Kutsky, Roman J. *Handbook of Vitamins and Hormones.* New York: Van Nostrand Reinhold, 1975.

Larkcom, Joy. *Vegetables from Small Gardens.* London: Faber and Faber, 1976.

Leighton, Ann. *American Gardens in the Eighteenth Century.* Boston: Houghton Mifflin Co., 1976.

Lovelock, Yann. *The Vegetable Book: An Unnatural History.* New York: St. Martin's Press, 1972.

Martin, Franklin W., and Ruberte, Ruth M. *Edible Leaves of the Tropics.* Mayaguez, Puerto Rico: Joint publication of the Agency for International Development, Department of State, and Agricultural Research Service, USDA, 1975.

McDonald, Lucille. *Garden Sass: The Story of Vegetables.* New York: Thomas Nelson, 1971.

Organ, John. *Rare Vegetables for Garden and Table.* London: Faber and Faber, 1960.

Pears Encyclopedia of Gardening: Fruits and Vegetables. New York: Pelham Books, 1973.

Pellegrini, Angelo M. *The Food-Lover's Garden.* New York: Alfred A. Knopf, 1970.

Recipes and Reminiscences of New Orleans. New Orleans: Ursuline Academy Cooperative Club and Ursuline Alumnae Assoc., 1971.

Robbins, Ann Roe. *25 Vegetables Anyone Can Grow.* New York: Dover Publications, 1974.

Scargall, Jeanne. *Pioneer Potpourri*. Toronto: Methuen Publications, 1974.

Schery, Robert W. *Plants For Man*. Englewood Cliffs, N.J.: Prentice-Hall, Inc., 1952.

Schindlmayr, A. *Useful Plants*. Translated and edited by Allan A. Jackson and Jean P. Jackson. London and New York: Thames and Hudson, 1957.

Schuler, Stanley. *Gardens Are For Eating*. New York: MacMillan, 1971.

Schwanitz, Franz. *The Origin of Cultivated Plants*. Cambridge, Mass.: Harvard University Press, 1966.

Schwartz, Florence, ed. *Vegetable Cooking of All Nations*. New York: Crown, 1973.

Simon, André L., and Howe, Robin. *Dictionary of Gastronomy*. New York: McGraw-Hill, 1970.

Tatum, Billy Joe. *Billy Joe Tatum's Wild Foods Cookbook and Field Guide*. New York: Workman Publishing Company, 1976.

Thompson, Homer C., Ph.D., and Kelly, William C., Ph.D. *Vegetable Crops*. 5th ed. New York: McGraw-Hill, 1957.

U.S.D.A. *Gardening For Food and Fun,* The Yearbook of Agriculture. 1977.

Vickers, Peggy. *The Free Food Cookbook*. Chatsworth, Calif.: Major Books, 1975.

Ward, Artemus. *The Encyclopedia of Food*. New York: Peter Smith, 1941.

Wythes, George, V.M.H., and Roberts, Harry. *The Book of Rarer Vegetables*. New York: John Lane Company, 1906.

Index

Flowering kale. *See* Kale, flowering
Fodder. *See* Animal feed
Foeniculum vulgaris dulce. *See* Florence fennel
Forcing, of chicory, 113-15
  of dandelion, 163
  of rhubarb, 335-36
  of sea kale, 355
French endive. *See* Chicory
French spinach. *See* Orach

G

Garden cress, 148-50
  cooking methods, 150
  nutritional value of, 150
Garden rocambole. *See* Egyptian onion
Garland chrysanthemum. *See* Shungiku
Garlic, 185-88
  cooking methods, 187
  flowering of, 187
  insect control with, 187
Girasole. See Jerusalem artichoke
Glycine max. See Soybean
Goa bean. *See* Asparagus pea
Goats, nettles as feed for, 259
Goatsbeard. *See* Salsify
Gobo. *See* Burdock
Goober. *See* Peanut

Good King Henry, 191-93
  cooking methods, 193
  nutritional value of, 193
Gooseberry, dwarf cape. *See* Ground cherry
Gorp, from peanuts and raisins, 300
Grain, amaranth for, 2-3
Green Chinese cabbage. *See* Michihli
Ground cherry, 193-97
  cooking methods, 196-97
Groundnut. *See* Peanut
Gumbo. *See* Okra

H

Haku-sai. *See* Chinese cabbage
Hamburg parsley, 283-85
  cooking methods, 285
  nutritional value of, 418
Healing herb. *See* Comfrey
Helianthus tuberosus. See Jerusalem artichoke
Hemerocallis. See Daylily
Herb patience. *See* Sorrel
Hibiscus esculentus. See Okra
Hinn choy. *See* Amaranth
Horse bean. *See* Fava bean
Horseradish, 199-202
  cooking methods, 202
  insect control with, 202

  medicinal value of, 200, 202
  varieties of, 199
Horticultural bean, 31-33
  cooking methods, 32-33
Husk tomato. *See* Ground cherry

I

India mustard. *See* Mustard
Indian cress. *See* Nasturtium, Purslane
Indian fig. *See* Prickly pear cactus
Insect control, companion planting for, 40, 118, 225
  garlic for, 187
  horseradish for, 202
  leeks for, 230
  screening for, 4
  See also individual pests
Interplanting. *See* Companion planting
Inulin, for diabetics, in Jerusalem artichoke, 208
  in scorzonera, 350
Ipomoea batatas. See Sweet potato
Italian celery. *See* Florence fennel
Italian dandelion. *See* Chicory

Italian pole bean. *See* Romano bean

J

Japan, use of wild vegetables in, 15
Japanese beetles, control of, 50
Japanese bunching onion. *See* Welsh onion
Japanese cucumber. *See* Chinese cucumber
Japanese eggplant, 173-76
    cooking methods, 176
Japanese leek. *See* Welsh onion
Jerusalem artichoke, 205-9
    cooking methods, 209
    nutritional value of, 208
    storage, 208-9
    varieties of, 205-6
Jicama, 211-14
    cooking methods, 213-14
    nutritional value of, 211, 214
    rotenone in, 214
John-go-to-bed-at-noon. *See* Salsify

K

Kale, 217-20
    cooking methods, 219-20
    nutritional value of, 218, 417
    varieties of, 218
Kale, flowering, 220-21
    nutritional value of, 221
Knitbone. *See* Comfrey
Knob celery. *See* Celeriac
Kohlrabi, 223-25
    cooking methods, 225
    nutritional value of, 224, 417
    storage of, 225
    varieties of, 224

L

Lactuca sativa asparagina. *See* Celtuce
Lady finger. *See* Okra
Lady Godiva pumpkin, 317-20
    nutritional value of, 317, 419
    varieties of, 318
Lamb's lettuce. *See* Corn salad
Land cress. *See* Winter cress
Lantern plant, Chinese, 195
Lathyrism, fava bean as cause of, 29
Leaf nutrient concentrate, of protein, 6-7
Leek, 227-32
    blanching of, 229
    cooking methods, 231-32
    insect control with, 230
    nutritional value of, 231, 418
    varieties of, 228
Leek, Japanese. *See* Welsh onion
Lemon lily. *See* Daylily
Lepidium sativum. See Garden cress
Libato. *See* Basella
Lincolnshire asparagus. *See* Good King Henry
Lotus tetragonolobus. See Asparagus pea
Love-lies-bleeding. *See* Amaranth
Luffa, 235-38
    cooking methods, 237
    harvesting, 237
    nutritional value of, 418

M

Maches. *See* Corn salad
Malabar nightshade. *See* Basella
Malabar spinach. *See* Basella
Manchurian squash. *See* Spaghetti squash
Mango squash. *See* Chayote
Manila bean. *See* Asparagus pea

Rocket, 339-41
    cooking methods, 340
Romano bean, 39-41
    cooking methods,
      40-41
Rootknot, control of, 105
Roquette. *See* Rocket
Rotenone, in jicama, 214
Rucola. *See* Rocket
Rumex. See Sorrel
Runner bean. *See* Scarlet
    runner bean

S

Sabadilla dust, as stinkbug
    control, 143
Salad burnet, 73-75
    cooking methods, 75
Salsify, 343-47
    cooking methods,
      345-46
    varieties of, 344
Salt bush. *See* Orach
Sanguisorba minor. See
    Salad burnet
Sanjaku cucumber. *See*
    Chinese cucumber
Scarlet runner bean, 43-45
    cooking methods, 45
    nutritional value of,
      415
Scorzonera, 349-51
    as carrot fly control,
      350
    cooking methods,
      350-51

nutritional value of,
    350
Screening, for insect con-
    trol, 4
Scurvy grass. *See* Winter
    cress
Sea colewort. *See* Sea kale
Sea kale, 353-55
    blanching of, 354
    cooking methods, 355
    forcing of, 355
Sechium edule. See Chayote
Shallot, 357-59
    cooking methods, 359
    varieties of, 357
Sheep's sorrel. *See* Sorrel
Shell bean. *See* Fava bean,
    Horticultural bean
Shungiku, 361-62
    cooking methods, 362
    nutritional value of,
      362
Siew choy. *See* Chinese
    cabbage
Sinus passages, cleared by
    horseradish, 200
Sium sisarum. See Skirret
Skirret, 365-67
    cooking methods,
      367
Slender nettle. *See* Nettle
Slugs, control of, 44
Snow pea. *See* Sugar pea
Solanum melongena. See
    Japanese eggplant
Sorrel, 369-71
    cooking methods,
      370-71

nutritional value of,
    370, 419
Soup greens. *See* Hamburg
    parsley
Sour sauce. *See* Sorrel
Southern peas. *See* Cow-
    peas
Soybean, 47-52
    as coffee substitute, 52
    cooking methods,
      51-52
    nutritional value of,
      51, 415
    varieties of, 48-49
Spaghetti squash, 377-79
    cooking methods,
      379
Spinach, Ceylon. *See*
    Basella
Spinach, Chinese. *See*
    Amaranth
Spinach, climbing. *See*
    Basella
Spinach, French. *See*
    Orach
Spinach, Malabar. *See*
    Basella
Spinach, New Zealand. *See*
    New Zealand spinach
Sponge gourd. *See* Luffa
Spring cress. *See* Winter
    cress
Squash borers, control of,
    319, 378
Stem lettuce. *See* Celtuce
Stinging nettle. *See* Nettle
Stinkbugs, control of, 143,
    265, 319

Winter cress, 147-48
  cooking methods, 148
Winter radish, 327-31
  cooking methods, 330
  varieties of, 327-28
Wong bok. *See* Chinese
  cabbage
Wood ashes, as pest con-
  trol, 105, 378

Wren's egg beans. *See* Horticultural bean

Y

Yam bean. *See* Jicama
Yard-long bean. *See* As-
  paragus bean
Yard-long cucumber. *See*
  Chinese cucumber

Yellow rocket. *See* Winter
  cress
Yuen tsai. *See* Coriander

Z

Zea mays everta. See Popcorn
Zoale, made from
  amaranth, 2